James Craig Watson

A Popular Treatise on Comets

James Craig Watson

A Popular Treatise on Comets

ISBN/EAN: 9783744678933

Printed in Europe, USA, Canada, Australia, Japan

Cover: Foto ©berggeist007 / pixelio.de

More available books at **www.hansebooks.com**

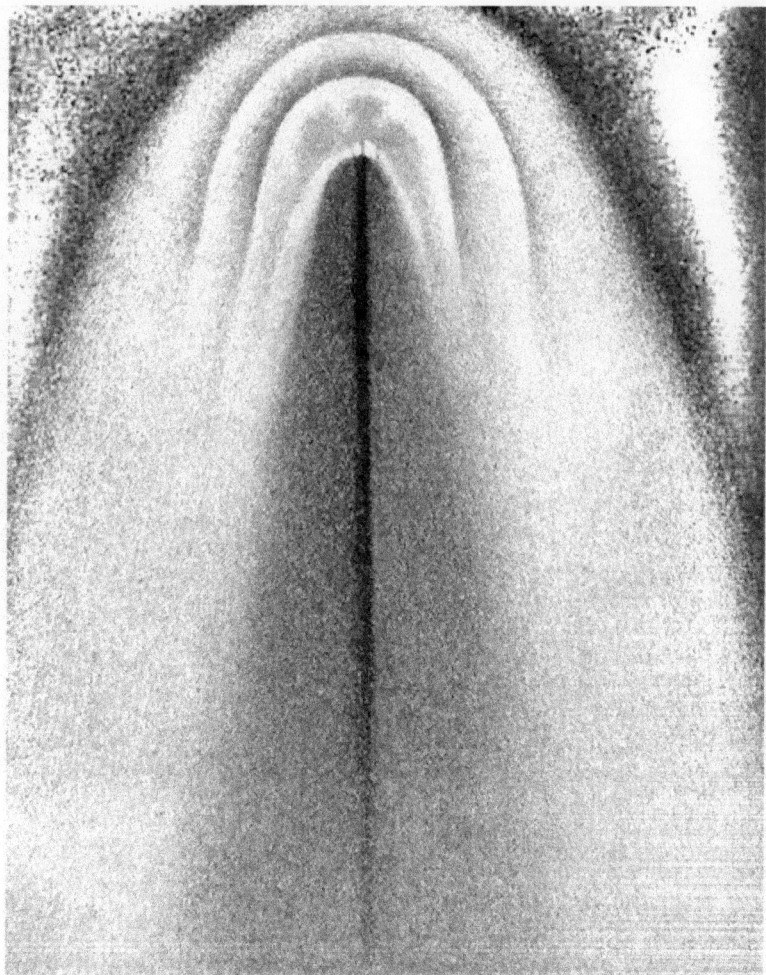

TELESCOPIC VIEW OF DONATI'S COMET OCT. 15, 1858.

A POPULAR

TREATISE ON COMETS.

BY

JAMES C. WATSON, A.M.,

FORMERLY PROFESSOR OF ASTRONOMY, AND NOW PROFESSOR OF PHYSICS, IN
THE UNIVERSITY OF MICHIGAN.

ERIT QUI DEMONSTRET ALIQUANDO, IN QUIBUS COMETÆ PARTIBUS CURRANT,
CUM TAM SEDUCTI A CETERIS ERRENT, ET QUANTI QUALESQUE SINT.
Seneca, Quæst. Nat., lib. vii., c. 27.

PHILADELPHIA:

JAMES CHALLEN & SON.

DETROIT: RAYMOND & LAPHAM.   CHICAGO: S. C. GRIGGS & CO.
ANN ARBOR: SCHOFF & MILLER.

1861.

TO

E. B. WARD, Esq.,

OF DETROIT,

AS A TESTIMONIAL OF THE LIVELY INTEREST WHICH HE HAS
TAKEN IN THE PROMOTION OF THE STUDY OF THE
PHYSICAL SCIENCES, THIS WORK IS MOST
RESPECTFULLY INSCRIBED, BY THE

AUTHOR.

PREFACE.

THE announcement of the discovery of a comet has always been received with great interest by both the learned and unlearned. By the former, it is hailed with delight, as affording new proofs of the perfection of the laws which regulate the motions of the celestial bodies; while to the latter, the appearance of a comet of any considerable magnitude furnishes, too often, the basis of the most foolish and superstitious fears, being regarded as the forerunner of pestilence, famine, or some other dreadful calamity.

The time, however, has now come when all such fears should be at an end, and it has been for the purpose of disseminating correct notions in regard to the nature and physical constitution of these wonderful bodies, among those not versed in astronomical science, that this volume has been prepared. The subject is here treated in the most complete and candid manner, and all technicalities are studiously avoided. In a word, the object has been to convey to every intelligent reader all the knowledge in regard to the motions and character of these erratic bodies possessed by the professional astronomer of the present day, without making it necessary to introduce the symbolic operations of mathematical analysis.

To accomplish this, it has been necessary in some instances to state facts without giving the processes which astronomers have employed in their determination; but the statements which have been made may be relied on as being accurate. Whenever the strictly popular character of the work has permitted a satisfactory explanation of the practical, as well as the theoretical, details of the calculations, it has been attempted.

Although it may be said that the age in which we live is indeed distinguished for the clearest and most enlarged views of social and political science, yet it is not less marked by the disposition, so unequivocally and universally manifested, to reject the inordinate estimate heretofore placed upon merely ornamental literature; and while it does not refuse their just rank and influence to such studies, it admits of that high consideration to which they are entitled, the sciences which explain the beautiful phenomena of the physical world. It is therefore encouraging to know that the demand for some portion of scientific knowledge, and the desire to be informed of what is passing in that universe of which our earth is so minute and apparently insignificant a member, no longer confined within the sacred precincts of astronomical observatories and academies of science, has spread throughout the whole extent of civilized society. To supply wants thus created, so far as they relate to those beautiful and mysterious worlds known as comets, this book is now offered with the hope that it may merit the approbation of an enlightened public.

J. C. W.

ANN ARBOR, MICHIGAN, *September*, 1860.

CONTENTS.

CHAPTER I.

CHAPTER II.

CHAPTER III.

CHAPTER IV.

CHAPTER V.

A

POPULAR TREATISE

ON

COMETS.

CHAPTER I.

INTRODUCTORY REMARKS — CHARACTERISTIC DIFFERENCE BETWEEN
COMETS AND PLANETS — SUPERSTITIOUS FEARS EXCITED BY THE AP-
PEARANCE OF COMETS — OPINIONS HELD BY THE ANCIENTS — FEARS
EXCITED IN THE MIDDLE AGES AND AT THE PRESENT DAY — INFLU-
ENCE OF COMETS ON THE WEATHER — THE GREAT DRY FOGS OF 1783
AND 1831 — POSSIBILITY OF A COLLISION BETWEEN A COMET AND
THE EARTH — EFFECT OF SUCH A COLLISION — TRUE THEORY OF
COMETARY INFLUENCE.

THERE is no department of human learning and
research in which the mind has been exercised with
more transcendant success than in the theory of
astronomy. The development of the legitimate
results of the theory of gravitation, first discovered
by Newton, and its extension to the far distant
regions of the Universe, has enabled the astronomer
of the present day to comprehend, at one glance, as
it were, the past, the present, and the future changes
of entire systems. It has enabled him to show that

2 (13)

even in our own solar system, with its planets and
their satellites, where everything is in a state of
motion apparently the most complex that could be
devised, although the mutual attraction of the
bodies which compose it, causes the orbits alternately to contract and expand, and their planes to
rock slowly up and down, yet, that stability is there.
It permits him also to declare that this unseen agent
of the Supreme Intelligence, mysterious in its nature
as spirit itself, connects the parts of the universe so
intimately, that action is answered instantaneously
by reaction through distances which elude even the
grasp of the imagination; and yet the law of this
force, though the most general and exalted that man
has discovered, is so simple that the effects of gravitation, however numerous and complicated, have
been, or may be, predicted with unerring certainty.
Thus do we find that as far as telescopic vision can
extend, in the most remote realms of space, sun
revolves around sun, and system around system, in
obedience to the same power that causes the rains
to descend and the tides to flow.

If we go back to the earliest ages of the world,
and trace forward the progress of astronomical discovery until the time of Newton, we shall find that,
notwithstanding the perseverance and acuteness of
the many illustrious astronomers who preceded him,
there were many phenomena presented in the hea-

vens which, with their knowledge of laws and forces, baffled every attempt at explanation, but which now present us only with striking proofs of the operations of the attraction of gravitation. The planets, which were never so distant as to escape vision, having perpetually afforded opportunities for observation, were known to move in obedience to some definite law of nature, no matter what that law might be. But there was another class of celestial bodies whose motions seemed to be at variance with any force which could harmonize with the actual motions of the planets; and these, from the circumstance of their having presented, in most instances, the phe- nomenon of a bright head or nucleus, with a train of fainter light bearing some resemblance to a tuft or lock of hair, were called comets (from the Latin word *coma*, signifying a bunch of hair), in order to distinguish them from the other luminaries, which, whether near or remote, apparently fixed or mova- ble, never exhibited any similar appearance. The planets, too, were known to move regularly, in their annual course, from west to east, in paths which were nearly circular, and which were confined to a very narrow belt or zone of the heavens; and con- sequently they were always visible, except when in the immediate vicinity of the sun. But the case presented by the comets was far different. They were seen only for a brief period, and were observed to move over a wide extent of the heavens, and in every direction with reference to the path pursued by the earth. Some were found to move in orbits whose planes must be nearly at right angles with the orbits of the planets, and others were found to move nearly in their plane. Some were seen to move from west to east, in the direction of the planetary motions, while an equal number retrograded, or from east to west; and from the appearances which they presented from day to day, it was evident that they approached very near the sun, and then retreated to the most remote regions of space.

It was without doubt this mystery which attended their motions, as well as that of their physical character, which has in all ages produced that sensation which we shall presently notice. The planets being always present, — before the invention of the telescope, — attracted no very considerable attention; but the comets, by their unusual aspect, their sudden arrival, and the prodigious velocity with which they winged their flight through our system into the deep recesses of the heavens, excited the most vivid impressions on the imagination. We shall find, therefore, that, since of all the celestial phenomena they are indeed the most wonderful, and especially calculated to excite in the minds of the ignorant the idea of supernatural agency, they have been universally regarded with the greatest apprehension.

Moreover, when we consider the wonderful characteristics which mark their flight through space, the suddenness and frequency of their appearance, and the incongruity of their motion towards the sun from all regions and in all directions, it will not perhaps seem strange that, in ages of ignorance and superstition, they have been regarded as the demonstrations of the wrath and as the harbingers of the vengeance of offended deities; and that, among the dreadful evils which they were supposed to forerun, were pestilence, famine, war, and political and physical convulsions.

It has been well remarked, that it is an inherent attribute of the human mind to experience fear, and not hope or joy, at the aspect of that which is unexpected and extraordinary. Thus it is that the strange form of a large comet, its faint nebulous light, and its unpredicted appearance in the vault of heaven, have in all countries been almost invariably regarded by the people at large as some new and formidable agent inimical to the existing state of things. The sudden occurrence and short duration of the phenomena lead to a belief in some equally rapid and portentous reflection of its agency in terrestrial affairs; while, on account of the almost infinite variety of the phenomena here presented, it is easy to find events that may be regarded as the fulfilment of the evil foretold by the appearance of

2 *   B

these mysterious bodies. Accordingly, we find, by reference to the early history of nations, when their philosophy was essentially ideal and superstitious, every celestial phenomenon which seemed to them to be extraordinary and at variance with the general harmony of nature, was regarded as indicative of some peculiar state of the weather, or of some event connected with the characters and fortunes of individuals. In climates where the sky was in a state of almost perpetual serenity, the magnificence of the starry heavens could not fail to excite the wonder and admiration of the inhabitants. Attracted thus unconsciously to mark the aspects of the heavenly bodies, they would, by continuous observation, soon acquire sufficient familiarity with the general appearance of the celestial vault, to detect any unusual appearance which might there be exhibited. While thus occupied in observing phenomena, they would necessarily endeavor to seek out causes and effects; and, since it is inherent in the human mind to attempt to pry into the future, it may readily be perceived that their methods of reasoning would be vitiated by labored attempts to connect the aspects of the heavens with terrestrial events. Speculating further, and reasoning by analogy, they would necessarily ascribe motion to the impulse of a living being, and as the simplest solution of every such appearance, they would suppose life to be inherent

in the body which moves. There is no language
which does not bear the traces of this belief. Every
fountain and every stream is at first conceived to be
"living water," every motion to be the breathing
of a spirit, and the clouds floating on high to be
borne along on "the wings of the wind." In this
way, therefore, they would finally arrive at the conclusion, whether legitimate or not, that the heavenly
bodies which were found to be in motion, must be
in a state of animation; and hence it would be easy
to imagine that their power must be incomparably
superior to any that could be produced by human
agency. When such an idea gains admission, the
imagination knows not where to stop; and it has
been asserted, as a natural sequence, that this ideal
and visionary philosophy would go still further, and
maintain that sympathies are established between
the most remote parts of the universe. All the divi-
sions of nature would be found to harmonize and
reciprocally actuate each other. The stars may
affect the earth, but the earth also affects the stars;
and hence it would be contended, that it is possible
to acquire such a knowledge of their mutual actions,
as not only to predict the changes which the former
may produce in the latter, but, to a certain extent,
to regulate the operations of the cause and to modify
the degree of the effect.

Such we conceive to be the primitive tendency

of the mind of man in the early and barbarous state
of society, in the study of astronomical science; but
in more advanced stages of civilization, the opinion
that the heavenly bodies have an influence over
terrestrial affairs, might appear to be confirmed by
observations and inductions which are not essentially
of so visionary a character.   Certain appearances in
the heavens being associated with corresponding
changes in the seasons, might be supposed to be
either the efficient causes or the invariable signs of
these changes.   The variations in the temperature
and density of the atmosphere, the ebbing and
flowing of the tides of the sea, and the fertility of
the earth, were known to bear some intimate relation to the sun and moon; and, consequently, it was
very naturally concluded that the planets and the
comets must have as great an influence over the
bodies and minds, the actions and fortunes of men,
as those more conspicuous luminaries have over the
vast realms of the ocean, the air, and the earth.
The myriads of stars which were seen scattered
here and there in bright profusion over the entire
surface of the heavens, would, on account of their
great number, be supposed to have been appointed
to regulate the destinies of the numberless individuals who inhabit the earth, to each of whom a
particular star was appointed, as the guide of his
conduct and as the arbiter of his fate; and, since

many of the stars were not observed to have any apparent connection with the great changes which have from time to time taken place in human affairs, it would be supposed that it was their exclusive province to preside over the incidents which occur in the minuter portions of the world. But the comets, from their sudden and yet casual appearance, and their hideous aspect, would be regarded as announcing those greater convulsions which have disturbed the nations of the earth. This opinion would be strengthened by the fact that it was generally admitted that there existed some intimate relation between mind and matter, and that the state of the human body and that of the mind are very closely connected. Now, since the bodily constitution is sensibly affected by the modifications of the atmosphere, it was very naturally concluded that the heavenly bodies might, through the medium of the atmosphere, affect the human body, and also, through the intervention of the body, affect the disposition and passions of the mind. It is well known that climate has a most powerful agency in the formation of human character, and consequently, if a variation of a few degrees in the mean temperature of any locality is sufficient to account for the different varieties of intellectual capacity, for the strength or weakness of passion, for the liveliness or defect of imagination, and for the activity or torpor of all the faculties, it would not be unreasonable to suppose that these varieties could be ascribed to influences from the celestial regions. Such a conception seems to harmonize completely with the doctrines advanced by some of the ancient philosophers, of the existence of two separate and distinct kinds of influence, — the one immediate and the other remote, the one discoverable by the senses, the other eluding the most careful observation.

We have thus discussed, somewhat in detail, what seems to be the only reasonable manner of accounting for the superstitious notions connected with unusual celestial appearances, which have prevailed in every age of the world and in every nation. We have been specific in the statement of these tendencies of the mind, in order that, although we may be amused at the recital of the follies and absurdities of those who fostered such delusive superstitions, we may still look beyond the immediate picture presented, and in the dim vista, it may be, witness the development of each peculiar system from what might be considered perfectly legitimate assumptions. With these preliminary remarks, we are prepared to consider the various opinions in relation to the appearances of comets, which have been entertained in each successive age of the world, commencing with that epoch, where the obscurity of tradition is dissipated by the light of

authentic history, and tracing them forward, step
by step, until the present day.

The ancients for the most part regarded comets
as simple meteors, generated by inflammable vapors
in the earth's atmosphere; and supposed that when
they were once extinguished, they were lost for ever.
Some, however, believed them to be distinct celestial bodies, situated beyond the moon. Diodorus
informs us that the shepherd astronomers of Chaldea
considered the comets as subject to the same dynamical laws as the planetary bodies, but revolving in
orbits which receded to a greater distance from the
earth. The ancient Chinese, unlike the Chaldeans,
were satisfied with simply extended observations of
comets, and did not attempt, even in the slightest
degree, the process of generalization and of deriving
any knowledge in regard to their nature and physical peculiarities.

Epigenes, who had studied among the Chaldeans,
very singularly maintained that there were two
kinds of comets, the one stationary and pouring
forth their heat in all directions, the other diffused
like hair and traversing the stars like the planets.
There were many also who believed that comets
were formed by a conjunction of two planets, the
light of both being confused into one, exhibiting
the phenomenon of an elongated star; or, that being
very near each other, the atmosphere (which they

supposed to pervade all space) was enlightened by both, producing the appearance of a comet. Apollonius Myndius maintained that a comet, instead of being composed of one or more planets or wandering stars in conjunction, was itself the same as a planet, although its form, unlike that of the other planets, was not globular, but more extended. He did not, however, suppose them to have any regular course, but to wander here and there, at random, throughout the entire heavens. The same opinion was entertained by Zeno the Stoic. He supposed that the comets have their regular courses, and that they reappear after the lapse of a very long interval of time. The majority of the Stoics, however, imagined the comets to be created in the atmosphere, and accounted for their motion by supposing that the rapid combustion which they believed to take place, impelled them forward in order to obtain a sufficient supply of fresh air.

Aristotle, in an attempt to explain the theory of the tails of comets, was led to believe that there existed some intimate relation between the comets and the Milky Way. He supposed that the myriads of stars which compose this starry zone, give out a self-luminous, incandescent matter. He therefore regarded the nebulous belt which separates the different portions of the vault of heaven as a large comet, the matter of which it is composed being

incessantly changing, by assimilating new particles
and giving off others. He also regarded comets as
presaging violent storms of wind and rain; and, on
account of their slow motion, declared that they
must be of immense weight. The remark has
indeed been very appropriately made, that, since
Aristotle exercised so great an influence throughout
the Middle Ages, it is very much to be regretted
that he was so averse to those grander views of the
elder Pythagoreans, which inculcated ideas so nearly
approximating to truth respecting the structure of
the Universe. He asserts that comets are transitory
meteors, belonging to our atmosphere, in the very
book in which he cites the opinion of the Pytha-
gorean school, according to which these cosmical
bodies are supposed to be planets, having long
periods of revolution. Panetius supposed the comets
to be delusive appearances, produced by certain un-
explained and indeterminate conditions of the atmosphere; and denied that they exhibited, either in
their motions or appearances, any of the attributes
of material bodies.

Such are some of the many curious theories which
were advanced by the ancient philosophers, in order
to account for the phenomena presented by the
comets. We must not, however, omit to give the
opinions held by Seneca, who, among the Greeks
and Romans, stands preëminent for his very accu-

3

rate notions in regard to the real character of these mysterious bodies. In speaking of the various theories which prevailed at his day, he says: "I do not agree with our philosophers, for I do not think that a comet is a sudden fire, but that it is to be regarded among the eternal works of nature;" and afterwards, in recapitulating his own theories, he adds: "The time will come when some one shall show us in what regions the comets wander, why they are so different from the other celestial bodies, and what and how great they are. Let us, therefore, be content with what we know, and leave it to those who shall succeed us to discover facts which our day and generation will not be able to unfold." The arguments by which Seneca enforced his own individual opinions, and the variety of the phenomena which he cited in support of each hypothesis, were wholly insufficient, as we shall see, to disseminate and perpetuate correct ideas of the character of the cometary worlds. The philosophy of those who succeeded him was so impregnated with a kind of religious superstition, that no argument, however plausible or cogent, seemed sufficient to disabuse the minds of the people of those ideas of fear which the appearance of a comet invariably excited. The tendency of the human mind to connect terrestrial events with celestial phenomena, which we have already discussed somewhat in detail, together with the ignorance and stupidity with which a bigoted and infatuated priesthood announced the most absurd and ridiculous doctrines, with the mistaken prestige of Divine authority, had prepared the people for the adoption of the most ludicrous and superstitious fears. Thus, while we are prepared to excuse the folly of less enlightened nations, we cannot avoid contemplating, with mingled feelings of ridicule and amusement, the devices employed even at a late day, by nations professing to have arrived at an advanced stage of civilization, in order to frighten away comets and counteract their supposed pernicious and baneful influences. Although the recital of the fears produced by comets in various ages of the world, and of the concurrent events which they were supposed to presage, may not be considered essentially necessary to an understanding of the true theory of the influence of these bodies, yet we deem it proper to give a few instances at least, in order that, separated as we are from the influences which operated in those times, we may regard them in all their unmasked absurdity. We shall therefore proceed at once to give some examples.

We are informed by Seneca, on the authority of Calisthenes, that immediately prior to the great earthquake in Achaia, in the year 373 B. C., there appeared a comet of enormous size, which was the

specific cause of the destruction of the cities of
Bura and Helice. Those cities were submerged by
the sea, and the appearance of the comet at the
same epoch, and its disappearance very soon after
that event, led to a very general belief that its office
was directly connected with the destruction of both
cities. The earthquake here referred to was, perhaps, the greatest and most disastrous which had
ever been felt in that portion of the globe. It is
not strange, therefore, that Aristotle should have
shared the general belief; and, accordingly, we find
that he makes mention of it as presaging the dreadful events which took place in the same year in
which it appeared. He informs us, also, that in the
year 468 B. C., there appeared a large comet, which
was supposed to have some agency in producing the
famous fall of aërolites near Ægos Potamos. One
of these meteoric stones, celebrated in antiquity, is
said to have weighed nearly two tons. It was seen
to fall from an immense cloud of fire and smoke;
and the occurrence of such an unusual event, together with the phenomena presented by the comet,
produced a most profound sensation among all
classes. Aristotle mentions another comet which
appeared in the year 341 B. C., which was supposed
to have caused a terrific storm near Corinth.

In the year 43 B. C. a comet made its appearance
which was so brilliant as to have been visible to the

naked eye in the daytime, when in the immediate vicinity of the sun. It was regarded by the Romans as the soul of Julius Cæsar — who had lately been murdered — transferred to the heavens. The star of the Magi, which signalized the birth of Christ, is numbered among the comets by many modern writers on the subject. The probability of the identity of the star of the Magi and a comet which appeared the same year, has been doubted. The comet is described as producing a light rivalling the splendor of the noonday sun. It is said to have appeared during an interval of twenty-four days, and its magnitude was so great that it extended over nearly fifty degrees of the heavens, so as to occupy more than three hours in rising and setting.

A comet appeared in the year 54 of our era, which was supposed to have presaged the death of the Emperor Claudius, which took place on the 13th of October of the same year; and a comet which appeared in the year 63 was supposed to have been the cause of earthquakes in Achaia and Macedonia. Josephus informs us, that just before the destruction of Jerusalem by Titus Vespasian, in the year 69, there appeared, hanging over the city, a flaming torch; and ten years later, while Vespasian was suffering from the disease of which he died, a comet was seen, which continued in sight during a long time. In the latter part of the year 190, or in the

3 *

beginning of 191, a "hairy star" was seen, which
Herodian informs us was the cause of many prodi-
gies which appeared at the same time. Stars were
continually seen by daylight, and some of them appeared to be stretched out lengthways, and seemed
to be suspended in the air.

The death of the Emperor Constantine is said to
have been preceded for several days by the appearance of a comet of extraordinary magnitude, and
another comet which appeared in the year 400 was
believed to have foreboded the most frightful disasters. We are informed by the historians of the
Western Empire that the misfortunes with which
Constantinople was threatened by Gainas, in the
same year, were so great that they were undoubtedly announced by the appearance of the comet,
which is described as being one of the most terrible
ones on record. It appeared above the city, with a
tail in the form of a sabre, extending from the highest region of the sky almost to the horizon. Two
years later another comet made its appearance,
which seemed fairly to rival this one. It was believed to have caused tempests and frequent thunder,
numerous eclipses of the moon, prodigious hail
storms, and spontaneous conflagrations. It was
said to have been attended by birds of evil augury,
and to have announced the arrival of Alaric, King

of the Visigoths, in Italy, an event which caused
the greatest possible consternation at Rome.

A comet which appeared immediately before the
birth of Mohammed was subsequently supposed, by
the followers of the Prophet, to have announced his
birth, in a manner precisely similar—except in detail—to the star which appeared at the birth of
Christ.

A comet appeared in March, 1402, which was so
brilliant as to be distinctly visible at mid-day; and
a second comet, which appeared in June of the same
year, was visible several hours before sunset. This
comet was believed to presage the death of John
Galéas Visconti. That prince, being a believer in
astrology, had consulted the charlatans of his court
in reference to such an event, and the fright produced by the appearance of the comet may possibly
have contributed to the fulfillment of the prediction
of his death. Another conspicuous comet appeared
in 1532, which was also stated to have been visible
before sunset. It created a most profound sensation throughout France, Germany, and Northern
Italy, where it was considered as the announcement
of the death of Sforza II.

The celebrated comet of Halley, which we shall
subsequently refer to, performs its revolution around
the sun in a period of 75 years, and, consequently,
has been visible at many different epochs. It may

not be uninteresting, therefore, in order to show the discrepancies of all superstitious theories, to give, in a connected manner, all the various events which the same body has been supposed to announce at different visits to our immediate vicinity, although the fact of the periodicity of the comet was unknown to those who attributed to it such remarkable qualities. In 1005 the appearance of this comet was attended by a great famine; in 1080 by an earthquake; in 1155 by a cold winter and the failure of crops; in 1230 by rain and inundations (part of Friesland was overwhelmed, with 100,000 inhabitants); in 1304 by great drought, and intense cold in the following winter, succeeded by a pestilence; in 1380 by a still more destructive contagion; and in 1456 by wet weather, inundations, and earthquakes. At its appearance in 1456 the comet is represented as being of an "unheard of magnitude," and as having a tail which extended over sixty degrees of the heavens, or nearly from the zenith to the horizon. It was visible during the month of June, and spread terror throughout Europe. It was supposed to announce the future success of the Turks under Mohammed II., who was then engaged in the subjugation of the Christian nations. He had already taken Constantinople in part, had advanced his forces even as far as the walls of Vienna, and had thus struck terror into the whole

Christian world. Pope Calixtus II., terrified at the appalling spectacle presented by the comet, and aroused by what seemed to be the inevitable fate of Christianity, directed the thunders of the Church against the enemies of the faith, both terrestrial and celestial. It had become imperatively necessary that something should be done immediately to counteract the baneful influences which were operating both above and below; and, consequently, the Pope ordered public prayers, and issued a bull, in which he anathematized not only the Turks, but the comet; and in order to perpetuate this manifestation of the power of the Church, he ordained that the bells should be rung at noon, a custom which is still observed in various Catholic countries. But, as Pontécoulant has aptly remarked, neither the progress of the comet, nor the victorious arms of the Mohammedans, were arrested. The comet tranquilly proceeded in its orbit, passing through its appointed changes, regardless of the thunders of the Vatican, and the Turks under Mohammed established their principal mosque in the church of St. Sophia.

In the year 1531 the appearance of this comet is said to have been attended by great floods; in 1607 by extreme drought, followed by a most severe winter; in 1682 by floods and earthquakes; in 1759 by rains and storms, and slight earthquakes; while in 1835, the date of its last appearance, nothing, so far as we are aware, has been attributed to it, except, it may be, some local epidemic or unfavorable state of the weather.

It is related that a comet appeared in the year 590, to the presence and influence of which was ascribed a fearful epidemic, which prevailed in that year, in the crisis of which the patients were seized with violent paroxysms of sneezing, often followed by death. It has been asserted, also, that it became the custom, when these paroxysms manifested themselves, for the bystanders to address their benediction to the sufferer, exclaiming, " God bless you ;" and to this has been attributed the custom which is prevalent at this day, to address one who happens to sneeze with the same words.

In a work on the atmospherical origin of epidemic diseases, written by an English physician, about thirty years since, it is asserted that, since the Christian era, the most unhealthy periods have been precisely those in which some great comet appeared; that such appearances were accompanied by earthquakes, volcanic eruptions, and atmospheric commotions, while no comet has been observed during healthy periods. Among the many cometary influences enumerated are the following: Hot seasons and cold, tempests, earthquakes, volcanic eruptions, hail, rain, and snow, floods and droughts, famines, clouds of midges and locusts, the plague, dysentery, and influenza. Each affliction is assigned to its comet, whatever kingdom, city, or village, the famine, pestilence, or other visitation may have ravaged. In making thus, from year to year, a complete inventory of the misfortunes of this lower world, the question has very properly been asked, who would not have foreseen the impossibility of any comet approaching the earth, without finding some portion of its inhabitants suffering under some affliction ; and who would not have granted at once, what Lubienietski has written a large work to prove, that there never was a disaster without a comet, nor a comet without a disaster? These are only a few of the influences recorded in the work above referred to, but there is one in particular which must be regarded as the masterpiece of absurdity, namely, that the appearance of the great comet of 1668 produced a remarkable epidemic among the cats in Westphalia.

The great comet which appeared in 1811 is said to have produced some of the most remarkable results. Among others, it is asserted, in a standard English periodical, published a few years subsequent to the appearance of the comet, that its influence produced a mild winter, a moist spring, and a cold summer ; that there was not sufficient sunshine to ripen the fruits of the earth ; that, nevertheless (such was the cometic influence), the harvest was abundant, and some species of fruits, such as melons and figs, were not only plentiful, but of a delicious flavor; that wasps rarely appeared, and flies became blind and died early in the season; that in the neighborhood of London, numerous instances occurred of women bearing twins, and it even happened, in one instance, that the wife of a shoemaker in Whitechapel had four children at a birth! All these singular influences were assigned to this comet. In Germany the somewhat remarkable crop of grapes produced in that year was duly acknowledged to be the effect of the comet; and the wine manufactured therefrom, which received the appellation of the *comet wine*, obtained a popularity and demand which has rarely, if ever, been equalled.

The celebrated traveller Rüppel, in writing from Cairo on the 8th of October, 1835, observes that the Egyptians thought, that the comet then visible was the cause of the great earthquake which was felt in that country on the 21st of August of that year, and that the comet also exercised so malignant an influence over some of the lower animals, that horses and asses perished in great numbers. The truth was, he adds, that the poor animals died of starvation, their usual forage having failed in consequence of the insufficient inundation of the Nile.

Such are a few examples of the many curious and absurd notions of cometary influences which have prevailed at various periods in the history of the world. We might suppose that in these enlightened times, no such attributes would be assigned to these chaotic worlds, especially in countries where education is generally diffused among the people, and where the arts and sciences have been carried to a high state of perfection. We ought to expect to find, in these cases, that the appearance of the comet would be everywhere hailed with delight, as furnishing new proofs of the beautiful harmony which characterizes the motions of the heavenly bodies. But such is not generally the case. The same tendency of the human mind to connect terrestrial events with celestial phenomena, exhibits itself even when civilization, culture, and refinement appear to have attained the foremost rank. We even find that in the days of Shakspeare, the comets were regarded as being directly connected with the great, more especially the calamitous events of nations, and are thus introduced by the illustrious poet himself, in the lamentation which the Duke of Bedford makes over the bier of Henry V.:

"Comets, importing change of times and states,
Brandish your crystal tresses in the sky;
And with them scourge the bad revolting stars,
That have consented unto Henry's death."

4

Milton, too, though he lived in the days of Kepler
and Galileo, though he was imbued with all the
learning and philosophy of his time, and showed
that he was intimately acquainted with all the labors
of the philosophers who had preceded or were
cotemporaneous with him, does not scruple to call
in the aid of the malign power of comets, in order
to heighten his picture of Satan when preparing for
the combat:

> . "On the other side,
> Incensed with indignation, Satan stood,
> Unterrified, and like a comet burned,
> That fires the length of Ophiuchus huge
> In the Arctic sky, and from its horrid hair
> Shakes pestilence and war."

When such men as Milton and Shakspeare allude
to the comets in such a manner, even for the purpose of illustration, we will not be surprised to
know that sentiments similar to these were entertained by the people at large. But to come down
even to a later day, to the date of the last appearance of Halley's comet (1835), we shall find that
notions fully as absurd as those which we have
already noticed, were received in France and elsewhere. But a short time previous to the approach
of the comet to the earth and sun, Arago wrote as
follows: "I would have wished, for the honor of
modern philosophy, to be freed from the necessity
of taking serious notice of such absurdities; but I

have acquired personal knowledge that some refu-
tation of them is not useless, and that the advocates
of these influences have no inconsiderable number
of followers.   Listen, when you are present at one
of those brilliant assemblies, where you meet what
is called good society; listen to the talk, of which
the approaching comet furnishes the subject, and
then decide if we ought to boast of that diffusion
of knowledge, which so many declare to be the
characteristic feature of our times."

Those, however, who assign to the comets, not an
influence over the fortunes of nations and individuals, directly, but rather an influence over atmospheric changes, may perhaps do so with a greater
show of probability.   To maintain that the comets,
owing to their peculiar nature, may, to a certain limited extent, operate to produce some unusual state of
the weather, will not at first be regarded by a large
portion of even intelligent people, as really absurd.
There are numerous reasons, and even forcible ones,
in theory, which may be adduced in support of such
a hypothesis; but there are other considerations
which are of far greater importance, and have a
bearing in the opposite direction, which must not
be overlooked.   It would, indeed, be easy to show
at once, upon general physical principles, that there
is no reason whatever why a comet should exert
any influence, even in the very slightest degree,

over the temperature of our seasons, or in producing
either droughts or rains; but it will perhaps be
more satisfactory to refute it by showing that such
doctrines are not in conformity with observed facts.
This method of showing the utter fallacy of attributing to comets any direct influence other than
what is due to their infinitesimal attractive influence
in accordance with the law of universal gravitation,
is most simple and easy. The appearances and
motions of the comets have been recorded. The
average daily, monthly, and yearly temperatures of
the weather, the direction and velocity of the winds,
the number and character of the storms, and the
extent of the droughts, have also been accurately
observed and recorded. To ascertain, therefore,
whether the comets really exercised any influence
on the temperature of the seasons, it is only necessary to place in juxtaposition these two records, and
to examine whether there exists any general law or
analogy which is capable of exhibiting any corre-
spondence between them.

Arago has given a table, in which he has exhibited in one column the temperatures of the weather
at Paris for every year, from 1735 to 1831, inclusive;
and in juxtaposition with these he has stated the
number of comets which appeared, with their magnitude and general appearance. The result is that
no coincidence whatever is observable between the

temperatures and the number and appearance of comets. Sometimes we find that the years of greatest mean temperature were those in which several comets appeared; and, again, that they were those in which no comet appeared. For example, in 1737, although two comets appeared, the mean temperature was inferior to that of the preceding years, in which none appeared. The year 1765, in which no comet appeared, was hotter than the year 1766, when two comets appeared, one of which was remarkable for its splendor; the year 1775, when no comet appeared, was hotter than the year 1780, which was marked by the appearance of two comets. The temperature was still lower in 1785, in which two comets appeared; while, on the other hand, the temperature of the year 1781 was greater, which was likewise marked by the appearance of two comets.

This question of the supposed connection between the temperature and the appearance of comets, was still further discussed by Arago. He has given not only the general temperatures, but also a table of the years of greatest cold, of the years in which the Seine has been frozen over, and also of the years of greatest heat; and he has shown that the corresponding appearances of comets have been varied without any connection whatever with these vicissitudes of temperature. Thus, 1806 and 1811 were

4 *

both hot years — the first, however, hotter than the second; and yet the first had only one comet, while the second had two, one of which was the most brilliant, with perhaps a single exception, which has appeared during the present century. Again, the year 1826, in which five comets appeared, was not so hot as 1831, which was characterized by the appearance of only one comet. But it is, perhaps, unnecessary to enter into any further detail in regard to the thermal character of particular seasons as compared with cometary appearances, and we may therefore notice some of the more prominent results arrived at by Arago.

Of seventy-four years, forty-nine were signalized by the appearance of one or more comets, and twenty-five by their non-appearance. The mean temperature of the former years was found to be 51.°6, and that of the latter 50.°7, the difference, as will be perceived, being actually less than a degree. Again, of the forty-nine years in which comets appeared, a single one was observed in twenty-five of these years, and two or more in the remaining twenty-four years. Now, if these bodies produced any influence on the temperature, it might naturally be expected, or rather it would be a necessary consequence, that a difference would be found to exist between the mean temperature of the latter and of the former years. It was found, however, that the mean temperature of the twenty-five years in each of which a single comet appeared, was 51.°6, while that of the twenty-four years in which there were several comets, was 51.°4; the difference being only the fifth part of a degree, and even that being in a direction contrary to the hypothesis that the comets tend to augment rather than to diminish the temperature. It is noticeable, however, in these comparisons, that as a general result, more comets are seen in hot years than in cold ones, which may be explained by the fact that the hot years, generally giving a finer sky, are more favorable than the colder ones for the discovery of those objects which are usually too faint to be detected by the naked eye. In the case of the large comets the circumstance just noticed would not produce any effect, and accordingly we find that in the year 1843, the date of the appearance of one of the most magnificent comets on record, the average temperature was more than a degree less than it had been for several years previous. Again, the mean temperature of the years 1853 and 1858, in each of which there appeared large and brilliant comets, did not sensibly differ from the average temperature of a series of years. Another fact which might be noticed in this connection, is that experiment showed, in the case of the great comet of 1811, that the light emitted by the comet, when brought to a focus

by a large mirror, was not equal to a tenth part of
the light of the full moon. Moreover, it was found
to have no sensible effect on the blackened ball of
a thermometer, which was so delicate a test of variation of temperature that it would have indicated
the hundredth part of a degree. This alone shows
the total inefficiency of comets to influence our climate even in the slightest degree. In fine, the complete discussion of the cometary and thermal observations, continued through the entire period during
which authentic records of each have been kept,
fully establishes the conclusion that there exists no
foundation whatever for the popular opinion that
the comets influence the seasons.

There is another point of view in which the supposed agency of comets in producing atmospherical
changes has been presented, which requires reasoning of quite a different character from that which
we have already given. It is imagined that the
comet has a certain magnetic influence, of which
the law of intensity and of operation is unknown.
For example, it is now universally admitted that
there exists some direct connection between terrestrial magnetism and the solar spots, and in the
same way it is conceived that there may exist also
a direct connection between the comets and our
atmosphere. Those who have advanced such theories have generally adopted Oersted's hypothesis.

that light, heat, and electricity are produced in the
same manner, yet under different circumstances;
and that it is the ethereal fluid which serves for the
transmission of light by successive undulations,
which furnishes the agent for heat and electricity.
They conceive, therefore, that the comet is in a state
of electrical excitement, and that it operates by in-
duction through the ether, and affects the atmosphere. This is a general statement of the theory,
although its development by different individuals
has led to a great variety of absurd, and, for the
most part, conflicting results. The best, and perhaps the only argument necessary to refute these
doctrines, is the fact that observation fails to exhibit
any agreement or harmony in the phenomena which
could be reconciled with any recognized or supposed
law in physics. It fails in every instance to exhibit
the least connection between atmospherical changes
and cometary appearances; and in view of all this,
we cannot conclude otherwise than that the appearance of a comet is no more a prodigy, and has no
more influence upon the fate of men or of nations,
than the appearance of the moon or of any planet
or star in the firmament. In reality it cannot have
nearly so much; for the moon, by causing the tides,
affects both the atmosphere and the weather, and
thus indirectly it affects also the human body. The
comets, on the contrary, are so distant, and either

their motions are so rapid or their substance so rare,
that none of them have been found to have any
material action on such of the planets as they may
have approached very near, although the planets
have been found to have a very considerable influence on them.

We have thus shown conclusively that the general
opinion which prevails, even at the present day,
among all classes of people, that the appearance of
a comet has reference to, or rather causes, some
unusual state of the weather, is without the slightest
foundation, either in philosophy or in fact. The
arguments are such as to admit of no equivocation,
and every intelligent reader will at once perceive
how untenable and unworthy of confidence the
whole theory of cometary influence in terrestrial
affairs must be. There are, however, two other
cases connected with this branch of our subject,
which must not be passed by unnoticed, and which
we shall now consider, namely, the possibility of the
near approach of a comet so as to produce atmospheric disturbances and currents, and also, the
possibility of a collision between a comet and the
earth, and the effect of such a collision.

We shall hereafter have occasion to show that
some of the comets have trains which extend to
such an enormous distance, that were the earth in
the line of the axis of the tail it would certainly be

enveloped in it, since the length of the tail often exceeds the distance of the comet from the earth. Should it happen, therefore, that on the approach of a very large comet to its perihelion, or point nearest the sun, the earth should have a position such that the sun and comet would appear to be in conjunction, or nearly so, the inevitable result must be that the earth would be surrounded completely by the nebulous matter which forms the tail of the comet, but which we shall find to exist in a state of great attenuation. Thus, although the possibility of such an event is fairly established, yet, if we reason from the theory of probabilities, and introduce the comets which have hitherto been observed as the basis of our computations, we shall find that the probability of an approach sufficiently near to envelop the earth in the tail, will not exceed the ratio of one to one hundred millions. In other words, let us suppose the number of comets which shall pass through our system in a given time, and whose perihelia are within the orbit of the earth, to be one hundred millions; then it is barely possible, or rather probable, that one of these may envelop the earth in its tail. When such an event will take place, whether it ever has taken place, or whether it ever will occur, is a problem which defies the power of human analysis: but since the possibility

cannot be doubted, it will not be improper to consider what would be the probable effect produced.

It is well known, that since the comets are found
to be composed of matter extremely attenuated, the
tails will, as a necessary consequence, be even more
attenuated; and in case the earth should be enveloped in one of these, it would hardly be expected
that the nebulous matter would even penetrate our
atmosphere to any considerable distance. There
might, however, exist such conditions, as regards
the nebulosity of the tail and the atmosphere of the
earth, as would result in the complete permeation
of the atmosphere by the matter of the tail. As a
general thing, no such result need be feared, in case
the circumstances were such as to render it possible;
and even if it should really happen, the effect produced might be most beneficial to mankind. The
nebulous matter might be of such a character as to
cause the most fatal epidemics; or, on account of
atmospheric disturbances and currents which it
would generate, it might, on the contrary, be such
as to produce a sanitary effect. That atmospheric
currents exist in certain parts of the globe, which
do exercise a healing influence over various kinds
of diseases, has been certainly established. For
example, the influence of the Harmattan, a periodical wind, which blows three or four times a year
from the interior of the African continent towards

the Atlantic coast, between the latitudes fifteen
degrees north and one degree south, is thus described by an English traveller:

" The periods of its prevalence are stated to be
chiefly from the end of November to the beginning
of April, its direction varying from east-south-east
to north-north-east. Its duration at any one time
varies from one to six days, and its force is always
very moderate. A fog, thick enough to render the
disc of the sun red, always accompanies this wind.
The particles deposited by this fog upon the leaves
of vegetables, and on the black skin of the natives,
appears always white, but the nature of this whitish
matter was not ascertained. It was remarked that
this fog was speedily dissipated by the sea; for,
although the wind was sensible on the sea at many
leagues from the coast, the fog became rapidly less
dense, and, at the distance of a little more than a
league, it disappeared.

" One of the characteristics of this wind and fog
is extreme dryness. When it continued for any
time, the foliage of the orange and lemon trees exposed to it became shriveled and withered. So extreme is this dryness, that the covers of books, even
when closed, locked in chests, and enveloped in
linen cloth, were curved by it, just as if they had
been exposed to the heat of a strong fire. The
panels of doors, and frames of windows, and the

5　　　　　　D

furniture, were often cracked and broken by it. Its effects upon the human body were not less marked. The eyes, lips, and palate were parched and painful. If the wind continued unabated so long as four or five days, the face and hands grew pallid. The natives endeavored to counteract these effects by smearing their skin with grease."

This description may serve to give a very correct idea of the character of the Harmattan, which we conceive to be not unlike what might be expected in case the earth should be enveloped in the tail of a comet. From what is said of its effect on the human body, it might be conjectured that its final influence would be highly insalubrious; yet it is further related that it proved to have the extreme opposite quality. It was found that its first breath completely banished intermittent fevers. It is asserted, also, that those who had been enfeebled by the practice of excessive bleeding then prevalent there, soon recovered their strength; that epidemic and remittent fevers, which had a local prevalence, disappeared as if by enchantment; and that it rendered infection incommunicable, even when applied by artificial means, such as inoculation.

The immediate effect of the passage of the earth through the tail of a comet, would be to produce unusual disturbances of the atmosphere, partly by currents and partly by electrical action; and also to produce the most intense fogs. It has, therefore, been conjectured that the great dry fogs which spread over a large portion of the surface of the earth in the years 1783 and 1831, were produced by the near approach of a comet to the earth, and that the tail of the comet must have either passed over the earth or over a part of it.

The great dry fog of 1783 possessed several characteristics which have induced the belief in its cometary origin; and one of these is the fact, that it commenced on nearly the same day at places very distant from each other. It lasted nearly a month, and it was found that its position was not in the least affected by the winds, which proves that the atmosphere did not convey it over the regions in which it prevailed. It was found, also, that it prevailed equally at all accessible heights above the surface, and that it was as dense upon the summits of the Alps as upon the plains of France. It caused a general sensation throughout Europe, and presented the following phenomena: Its density was such, that in some places objects at the distance of a few rods could not be distinguished, and even such as could be distinguished appeared blue, or else surrounded with vapor. The sun appeared red, and without brilliancy, and could be gazed at with the eye unprotected at mid-day; while at its rising and setting, it disappeared in a sort of haze before

it had approached within ten or twelve degrees of
the horizon.

This fog was first seen at Copenhagen on May
29th, and was preceded by a succession of fine days.
In other places it was preceded by a gale.   In England it came after continuous rains; at La Rochelle,
it was seen on June 6th and 7th; at Dijon, on the
14th.   It was noticed almost everywhere in Germany, France, and Italy, from the 16th to the 18th;
on the 19th it was observed at Franecker and
throughout the Netherlands; on the 22d, at Spyd-
berg, in Norway; on the 23d, at St. Gothard and at
Buda; the 24th, at Stockholm; the 25th, at Moscow;
toward the end of June, in Syria; and, on the 1st
of July, in the Altai.   It will thus be perceived that
it covered a part of the earth's surface, extending
north and south, from Africa to Sweden; and we
might remark, in addition, that it prevailed on the
North American as well as upon the European
continent.

An examination of this fog showed that it was
absolutely dry, since the most delicate hygrometric
instruments exposed in it indicated the complete
absence of humidity.   It was found, also, to have a
faintly luminous quality, such as might be supposed
to proceed from a slight degree of phosphorescence;
and some observers maintained that they found
traces of acid in it.

In order to explain all these phenomena, which had never before been noticed in the case of the dry fogs which sometimes prevail in different parts of Germany, the cometary hypothesis, which referred its origin to the nebulous and slightly phosphorescent matter of the tail of a comet which may have been at that time passing over the earth, has been advanced. To decide this question definitely is a problem which admits of a ready and easy solution, and one which any one can fully comprehend. All the phenomena presented in the case of the fog, both in respect to time and place, we have already minutely described, and it remains only to see whether a comet could have occupied for so long a time a position such as the admission of the cometary hypothesis would require. The tails of comets, as we shall see hereafter, are usually in a direction opposite to the sun; and it results, therefore, that in order to satisfy the necessary conditions, the comet must have been almost exactly in conjunction with the sun. Now, for reasons which it would be useless to attempt to explain, but which may be readily conceived of, there is no possible combination of the motion of the comet in its orbit, and of the diurnal motion of the earth, which would be compatible with the position and continuance of the great dry fog of 1783. Had the interval during which the fog prevailed been much shorter, and the

5 *

date of its first appearance general, or nearly so,
the hypothesis of a cometary origin might be presented with a much greater show of probability;
but when we remember that the angular motion of
a comet, properly situated, cannot be reconciled
with the continuance of any such phenomenon for
a period longer than a very few days, we can conclude, in view of all the facts, with the utmost certainty, that the phenomenon under consideration
did not arise from the immersion of the earth in
the tail of an unseen comet.

The great fog of 1831 was, in every respect, excepting only the extent of country over which it
prevailed, similar to that of 1783. It appeared in
the month of August, commencing on the north
coast of Africa on the 3d, at Odessa on the 9th,
throughout France on the 10th, in the United States
on the 15th, and in China during the latter part of
the month. Observers in the north of Africa, in
the south of France, in the United States, and in
China, reported that the disc of the sun seen
through the fog had the tint of azure, and, in some
places, of emerald green. During the day the light
blue of the sky appeared to be tarnished by a mix-
ture of a dirty color. At an altitude of a few degrees above the horizon, the blue of the sky was
suddenly interrupted, and was terminated above by
a ring of dull brown-red color, more or less clearly

defined. Distant terrestrial objects of a deep color appeared to be nearly effaced, or covered by a bluish veil. It was found, also, that this fog had a proper light, similar to that described in connection with the great fog of 1783, and it was asserted that this apparently phosphorescent light was such that in the absence of the moon, even at midnight, it was possible to read the smallest written or printed characters.

Another fog of similar character appeared in the months of May, June, and July, 1834, which was confined almost exclusively to the western portion of Germany; and it is recorded, also, that similar fogs appeared in the years 526, 1721, and 1822. In each of these instances, however, as we have already noticed in the case of the fog of 1783, the hypothesis of the immersion of the earth in the tail of a comet is completely overthrown by the fact that the fog, although extensively spread, was not continuous, and certainly not uniform. Some parts of the European continent were altogether or nearly free from it, while in other parts it was developed in very different degrees. The times of its continuance in different places also varied much and irregularly, and in such a manner as to be entirely incompatible with the cometary hypothesis. Thus do we find that each new case presented furnishes additional evidence to show, that however mysterious and unac-

countable these phenomena may have been, yet that
we are of necessity compelled to refer them entirely
to terrestrial agency, and to admit that they had no
connection whatever with any celestial phenomenon.

At the times when these great fogs appeared, the
supposition of cometary agency became very prevalent; but in the inquiry which was excited, failing
to obtain any tangible evidence in support of such
a hypothesis, various other theories were devised.
Lalande attributed the fog of 1783 to the quantity
of electricity developed during a very hot summer
that succeeded a moist winter; while Cotte regarded
it as formed of metallic emanations united with
electricity, in consequence of the great heat and
numerous earthquakes which marked the summer
of 1783. It is a historical fact, that in the year in
which this fog appeared, there was a very violent
earthquake in Calabria, and a volcanic eruption in
Iceland, and very many have attributed the existence of this fog to these. It has been determined
by accurate researches that the column which rises
above a volcano, although bearing a very great
analogy to a column of smoke, is composed, for the
most part, of the vapor of water and of volcanic
ashes, with which are mixed transparent gases in
various quantities. When the lava runs over the
side of the mountain, it carbonizes everything
which it meets, and an immense cloud of smoke

rises in the air. Considering, therefore, the fact that an immense quantity of vegetables were consumed in Iceland, as well as seventeen villages, we can readily comprehend that the lava, when running over a soil covered with these vegetables, might have been able to produce this smoke, which the north winds, then prevalent, would immediately spread over a great portion of Europe. Add to this, that the combustions of turf and the conflagrations of forests were both extensive and frequent during the summer of 1783, which was a very dry one, and we can conceive of the distribution, in a very short space of time, of the vapor, smoke, and gaseous matter combined, over nearly the entire surface of the globe, and under conditions, as respects its general characteristics and electrical influence, producing all the phenomena which were observed.

Such are some of the most prominent examples of the supposed agency of comets in atmospherical disturbances, by permeating the atmosphere of the earth with their gaseous matter; and although, in each of these cases, this hypothesis has been found to be insufficient for the explanation of the phenomena, yet we cannot, on this account, venture the assertion, that absolutely no such event, as the cometary hypothesis under consideration would seem to indicate, can ever take place. We have already seen that there does exist a possibility of the passage of the tail of a comet over the surface of the earth, so as to envelop either the whole or a part of its surface; but we have also seen that the probability of such an occurrence is so small that it can hardly be expected; and in proof of this assertion we have the fact, that during the period of authentic history, there is no phenomenon on record which may not be satisfactorily explained without a resort to any other than terrestrial agency, which would otherwise be referred to a cometary origin.

Having thus disposed of cometary influences, so far as they have been supposed to relate to the birth and death of heroes, to the prevalence of famine, pestilence, war, and various other political and physical convulsions which have from time to time taken place in the affairs of the world; and having shown conclusively, as must be admitted, that the comets have no influence whatever over atmospheric changes, either directly or indirectly, we are prepared to consider the other question which we have proposed as relating further to this branch of our subject, namely, the possibility of a collision between a comet and the earth, and the probable effect of such a collision. The question thus presented is one which admits of the closest reasoning, and one in which the results arrived at, admitting the validity of the argument, must be morally certain.

We are here not left to the guidance of facts contradictory and ill authenticated; but, on the contrary, the data are both abundant and accurate.
Under such circumstances, the case presented cannot fail to be of the greatest interest, inasmuch as
it involves consequences which may ultimately prove
disastrous to the inhabitants of the earth, if not to
the existence of the planet itself in its present condition.

When we consider the great number of comets
that are known to be continually traversing our
system, and also the fact that the orbits of some of
them, as we shall subsequently see, almost intersect
the orbit of the earth, we may not hesitate to admit
at once the possibility of a collision between the
earth and one of these bodies.   The circumstances
under which such a catastrophe may happen — for
such it would certainly be — are these: The orbits
of the earth and comet must either actually intersect each other, or at least approach so near at some
point, that the distance between the orbits shall be
less than the sum of the radii of the nucleus or
head of the comet and the earth; and, the earth
and comet must both arrive at this point at nearly
the same instant.   These conditions are indispensable to a collision; but it would, indeed, be possible, without producing a collision, for an approach
of a comet to the earth to such an extent that its

orbit would be changed, or else it might be drawn to the earth, in case the attractive force of the latter exceeded that of the sun. Another indispensable condition to the possibility of a collision, would be that the earth and comet should move in contrary directions, or, in other words, that the motion of the comet should be retrograde. The planets all revolve around the sun from west to east, while the comets, as we shall notice, are not all found to move in the same direction. Some of them move from west to east, in the direction of the planetary motions, while perhaps an equal number move from east to west, or in a direction contrary to that of the planets. Should the earth come in contact with a direct comet, they would continue to move on in new orbits, which would necessarily result from their impinging; and the deviation in either case would be least when the planes of the two orbits were nearly coincident, and when they crossed each other very obliquely. Should the orbit, however, of a comet which moves with retrograde motion, as compared with the planets, intersect the orbit of the earth, and the comet and earth happened to meet at this point, the result would be an absolute collision, the effect of which would certainly be disastrous to the comet, if not to both bodies. Since, therefore, the conditions on which a collision depends are such as may actually exist, it may not be improper

to consider at once what must be the inevitable
result of such an event, before discussing the absolute probability of its occurrence.

We have just remarked, that in the case of a
simple appulse between a comet and the earth, or
in the case that the motions of the comet and the
earth should both be nearly in the same direction,
the effect produced would be merely local in its
character, and might perhaps cause a deflection in
the path of the earth, and change its velocity. If,
however, the comet had no solid nucleus of any
considerable size, the effect either of an appulse, or
even of a direct collision, would not probably be
different from that produced by the passage of the
tail of a comet over the earth, which we have already
noticed. The effect produced in this case would be
nothing more than an intermixture of the cometic
atmosphere with that of the earth. The extremely
light mass of the comet would, notwithstanding its
proximity, render it impossible that it could produce
any sensible effect, either on the annual or diurnal
motion of the earth; and, consequently, no change
would be produced in our days, seasons, and years.
But in the case where the comet had a solid nucleus,
and withal moved in a direction contrary to the
annual motion of the earth, the most fatal effects
would be produced by a collision. The momentum
might indeed be sufficient to destroy the progressive

6

motion of both bodies; and it would result from this that the attractive force of the sun would cause both to fall to his surface. Such would be the fate of the earth if it were struck by a comet, with a mass only about four times that of the moon, and moving with a retrograde motion at the rate of about three hundred and fifty miles in a second of time, and both bodies would fall to the sun in sixty-four and a half days. In case any obstruction had destroyed the centrifugal force of the great comet of 1680, when nearest the sun, it would have fallen to his surface in three minutes.

Let us, however, suppose a case which affords a greater degree of probability, — since the one just supposed is such as is almost, if not entirely impossible, — namely, that the shock experienced by the earth from a collision was not such as to destroy the centrifugal force. It can readily be understood, without further explanation, that the effect of the concussion in this case would be to change both the axis around which it performs its diurnal revolution, and the time of rotation. The fluids composing the atmosphere, the oceans, seas, lakes, and rivers, not partaking of that change, would be thrown from their position of relative equilibrium. Violent atmospheric commotions would ensue, and there would also be a sudden rush of all the waters of the ocean, lakes, and rivers from their former beds, overflowing the land, and sweeping before them animate and inanimate beings in one vast and un-distinguishable ruin. The entire animal, and perhaps the vegetable creation, would be destroyed by a universal deluge, or by the violence of the colli-sion. Every vestige of animated existence would be forever obliterated, and all the monuments of human industry overturned. It has, therefore, been asserted that in this way we may explain why the ocean has deserted the most lofty mountains, leav-ing, however, incontestible evidence of its presence; why the animals and plants of the tropics may have existed in the higher latitudes, where their relics and footsteps are still seen. It has further been asserted that this will explain the recent date of the present races of men, whose earliest monuments go back only a few thousand years; since the human race, reduced to a small group of individuals, in a deplorable condition, occupied exclusively in pro-viding for their physical wants, must necessarily have lost the remembrances and records of all the sciences and arts; and when, later, new wants were created by the progress of civilization, everything was to be recommenced, as if no previous progress had been made, and as if man had been for the first time placed upon the earth.

It is well known, that all parts of the globe bear testimony to the vast and destructive floods of mighty waters. The remains of plants and animals deeply buried in the ground, and the "footprints" in the solid rocks, declare that long before man became its inhabitant, this earth of ours was tenanted by innumerable races of beings altogether different from those which share with us the present state of things. These must have ceased to exist as animated beings thousands and thousands of years ago, since the strata under which they are found indicate that sometimes the waters have prevailed for many centuries over the beds containing them, and sometimes' the dry land. Thus, too, do we find that enormous masses of rock have been torn from their native hills, and have been carried to distant regions, and these, together with the deposits of the natives of the deep on the tops of the loftiest mountains, declare the irresistible force and magnitude of the vast waves which, in remote times, have carried destruction over the face of nature. These and similar effects have been attributed to the shock which may have been experienced from a comet in those remote ages of the world.

The appearance of the comet of 1680 led Whiston, the friend of Newton, and the man whom he had designed as his successor, to make an attempt to account for the Mosaic deluge by supposing the earth to have been struck by a comet at that time.

Not content with simply conjecturing the possibility of explaining the deluge in this manner, he attempted further to exhibit a coincidence between the theory and the circumstances of that great catastrophe, as related in Genesis.

The biblical deluge happened in the year 2349 before the commencement of our era, according to the modern Hebrew text; and in the year 2926, according to the Septuagint and Josephus. Whiston, therefore, endeavored to prove that a great comet may be supposed to have appeared at that time.

It is related, that a large comet appeared in the year 1106, which resembled the blaze of the sun, and had an immense train. In ascending still higher, we find a very large and terrific comet, designated by the Byzantine historians by the name of Lampadius, whose appearance must have been in the year 531 of our era. A comet also appeared in the month of September, of the year 43 before our era, shortly after the death of Julius Cæsar, which is said to have been very brilliant, and to have been visible for some time before sunset.

A comparison of these dates reveals the fact that, supposing these to have been successive appearances of the same comet,—which, it should be remarked, is altogether conjectural,—the period of revolution of the comet of 1680 differs but little from 575 years. Adopting this hypothesis, and computing

6 *   E

backward, we find that four revolutions prior to
B. C. 43, would give the epoch of the deluge, according to the modern Hebrew text, within six
years; and by supposing that there had been five
revolutions, we obtain the date of the deluge, according to the Septuagint, within eight years. From
these and other considerations, it has been inferred
that, as Whiston has supposed, there may have
been a comet near the earth at the period of the
deluge, although the probability of its having been
in collision admits of no such assumption. It ought
also to be remarked that the comet of 1680 moved
in the same direction, relatively, in which the earth
moves; yet the plane of its orbit was so inclined
to that of the earth's orbit, that a collision, although
barely possible, yet could not have been so disastrous as might have resulted in case the motion had
been retrograde.

But without entering into a detailed account of
the arguments adduced by Whiston in support of
his hypothesis, it will perhaps be sufficient to notice
only the general conclusions which he arrived at.

He asserts, that at the period of the deluge, the
comet of 1680 was only nine or ten thousand miles
from the earth; and, consequently, that it attracted
the waters from the great deep, just as the moon is
known to attract the waters of the ocean. Its action,
on account of that great proximity, must have

tended to produce an immense tide; and hence, he
contends, that since it is probable that the terrestrial
shell could not resist the impetuosity of the inundation, it would break in at a great number of
points, and the waters, then free, would spread
themselves over the continents. To render this
intelligible, it might be well to state that he supposed the earth to have been originally a comet.
He therefore regarded it as having, at or near
its centre, a nucleus, which is a hard and compact
substance, and which was formerly the nucleus or
head of the comet. He supposed, also, that the
matters of various natures, confusedly mixed, which
composed the nebulosity, subsided more or less
quickly, according to their specific gravities; that
the then solid nucleus was at first surrounded by a
dense and thick fluid; that the earthy matters pre-
cipitated themselves afterwards and formed a cover-
ing over the dense fluid — a kind of crust, which
may be compared to the shell of an egg; that the
water, in its turn, covered this solid crust; that in
a considerable degree it became filtered through the
fissures, and spread itself over the thick fluid; and
that, finally, the gaseous matters remaining suspended, purified themselves gradually, and constituted our atmosphere. Thus he makes it possible,
as he imagines, for the supposed shell to be broken
up by the rush of the exterior waters; and when

once broken up, the waters which he supposed to
have existed interior to this crust, and on which he
supposed it to rest, would gush forth, and produce
the phenomena given in Genesis, namely: "In the
second month, the seventeenth day of the month,
the same day were all the fountains of the great
deep broken up."

Continuing his theory still further, Whiston supposed that the nebulosity of the comet reached the
earth near Mount Ararat, and that the range of
mountains intercepted the entire tail. He concludes, therefore, that it would result, as a necessary
consequence, that the terrestrial atmosphere, thus
charged with an immense quantity of aqueous particles, was sufficient to produce forty days' rain of
such violence as the ordinary state of the globe can
give us no idea.   Thus he reconciles the remainder
of the passage found in Genesis: "And the windows of heaven were opened. And the rain was
upon the earth forty days and forty nights."

Such is the celebrated theory of Whiston in relation to the manner in which the Mosaic deluge was
accomplished; a theory which is remarkable from
the fact that it was advanced and maintained by a
man who was an intimate friend of the great Newton, and who must have been acquainted with the
labors of that distinguished philosopher. In many
points this theory may seem to present, at least, some

show of probability, but there are so many other points in which it presents obvious and glaring defects, that we can regard it as nothing more than the fancy of a mind devoted exclusively to similar speculations. Geologists have shown that the earth has indeed an exterior crust or shell, the thickness of which is very small compared with the diameter of the globe, but that the interior, instead of being in the state supposed by Whiston, is, on the contrary, in a state of intense heat, such, in fact, that the great bulk is one vast mass of melted matter. This militates strongly against the theory of Whiston; and were there not additional considerations to counterbalance, would be sufficient to show the utter fallacy of all his arguments in relation to the deluge.

In order to give a greater weight to his arguments, and at the same time to remove objections which would otherwise serve to lessen materially the strength of his theory, Whiston supposed that the mass of the comet was at least six times greater than that of the moon. It can, however, be readily shown that even this enormous magnitude of the comet would not be sufficient to produce any such tides as he supposes to have been caused; and that the comet could not, by any possibility, have remained more than two hours and a half so near the earth as a fourth part of the moon's distance. It

may also be shown that it could not have remained
even for that length of time, unless it passed the
earth under a very improbable and peculiar combination of circumstances. Under such circumstances,
the production of a tide would be impossible, since
eleven hours at least would be necessary to enable
a comet to produce an effect on the waters of the
earth, from which the injurious consequences of a
deluge could follow.

Thus we see how utterly untenable are the doctrines of Whiston in relation to the deluge. They
are, perhaps, not less romantic than another theory
of the same man, of the probable object of the
comets in the economy of nature. He fixed on
these erratic worlds the residence of the damned,
and thus determined the nature of their punish-
ment. According to his theory, a comet was the
awful prison-house in which, as it wheeled from the
remotest regions of darkness and cold into the very
vicinity of the sun, hurrying its wretched tenants
to the extremes of perishing cold and devouring
fire, the Almighty was to dispense the severities of
his justice.

Remarkable, then, as the theory of Whiston is
found to be, it is a fact well established, that it was
received with a great deal of favor at the time when
it was promulgated, and many others following in

his footsteps have modified it, and extended it to suit their own peculiar fancies.

If, however, we suppose the earth to have received a shock from a collision with a comet of any considerable magnitude, it would not be difficult to conceive of a deluge such as described in Genesis. Some writers have adopted this hypothesis, and have carried it still further, and suppose the floods of mighty waters which have left their mark everywhere on the surface of the globe, have been produced by the shock of a comet in former ages. But the astronomer has shown that such is not the case; that the length of the day, which is the measure of the celestial motions, is immutable, and exhibits no trace of change; and that, if the earth had ever been struck by a comet so as to change the axis about which it performs its diurnal rotation, the effects would still be perceptible in the variation which it would have occasioned in the geographical latitudes. It may be demonstrated, however, that a spheroid, such as the earth is known to be, cannot permanently revolve around any axis except its shortest diameter, in case it was placed in any resisting medium, such as our atmosphere. Now, if by the collision of a solid comet, the earth were made to revolve on any other diameter than its shortest, — which is the one around which observation proves that it now revolves, — it would seem, for a reason similar to that just noticed, that it could not continue so to revolve, and that it ought to change its axis from hour to hour, until at length it would again revolve round its shortest diameter. It is, however, certain that the axis of the earth's rotation has never been changed, because, as the ethereal fluid which is supposed to pervade space offers no sensible resistance to so dense a body as the earth, the libration would to this day be evident in the variation of geographical latitudes. As nothing is known of the primitive velocity of the earth, it is possible that a comet may have given it a shock without changing the axis of rotation, but only destroying a part of its tangential velocity, so as to diminish the size of the orbit. In this case the effect would have been to increase the temperature of the seasons; but geologists have shown, that instead of the earth having become warmer, the tropical nature of the fossil remains found in the most northern countries of Europe and America, declare that the general climate of the earth is of a lower temperature now, than it was in the extremely remote ages in which these plants and animals must have flourished.

We see, therefore, that the consequences of a collision between a comet and the earth might be frightful in the extreme; yet it may be demonstrated that there is but little chance that such a collision

can ever take place. Indeed, it has been found by
actual calculation, from the theory of probabilities,
that if the nucleus of a comet having a diameter
equal only to one fourth part of that of the earth,
should come nearer the sun than the earth, its orbit
being otherwise unknown, the probability of the
earth receiving a shock from it, is only one in two
hundred and eighty-one millions. It is found, also,
that the chance of our coming in contact with its
nebulosity is not more than ten or twelve times
greater. We may rest satisfied, therefore, that,
although it cannot be affirmed absolutely that the
earth will never come in collision with a comet, there
is no reasonable cause to dread such an event, since,
by supposing the number of comets which pass near
enough to the sun to satisfy the required conditions
to be three in each year, which is perhaps too large
an estimate, we arrive at the conclusion that the
earth may receive a shock from one of these bodies
in the course of the next ninety-four millions of
years, while the probability is nearly as great that
such an event will never occur. This result, although
to those unacquainted with the mathematical theory
of probabilities it may seem conjectural, is worthy
of the very highest confidence ; and we may unhesi-
tatingly declare that although, in strict geometrical
rigor, it is not physically impossible that a comet

7

should encounter the earth, yet the moral possibility of such an event is absolutely nothing.

The only instance in which any such serious catastrophe has been anticipated, to any considerable extent, was in the case of the return of the comet of Biela to its perihelion, or point nearest the sun, in the year 1832. A detailed account of this comet will be given in the proper connection; but it may be well to state here, that this appearance of the comet excited the liveliest interest throughout the civilized world. A short time before its appearance, Damoiseau created very serious apprehension in France and elsewhere, by predicting that the comet would pass within eighteen thousand four hundred and eighty-four miles of the earth's orbit, a little before midnight, on the 29th of October, 1832; and since Olbers had computed that the radius of the comet's nucleus or head — which is the distance from the centre of the comet to its surface — would be twenty-one thousand one hundred and twenty miles, it was evident that its nebulosity would envelop a portion of the earth's orbit, and if, by any cause, the comet should be retarded one month in its revolution, it would undoubtedly meet the earth at that point. This startling announcement, emanating from a source which could not be questioned, excited the popular apprehension to the very highest degree.

Arago, however, dispelled the fears of his coun-
trymen, by showing that the comet would not be
retarded in its course, but that it would pass nearest
to the earth, at a distance of fifty-five millions of
miles; and, consequently, could not have the
slightest influence in terrestrial affairs, since it had,
in 1805, been ten times nearer without any sensible
effect, and without exciting any alarm.

We have now completed the account of cometary
influences, both real and imaginary, and more especially the latter; and what has been said ought to
be sufficient to assure every intelligent person, that
every superstitious notion which may have been
attached to them, every baneful influence which
they may have been supposed to exert, every supposed influence in producing atmospheric disturbances, and in changing or modifying the temperature
of our climate, and further, that all fears of a collision between one of these bodies and our earth,
must be regarded as one of the aberrations of the
human mind. We have been able to trace, however imperfectly it may have been, some distant
resemblance at least between these notions of cometary influence and the development of the philosophy of a people; and we have noticed that there
is even a tendency in the human mind to cling to
all such ideas, no matter how absurd, after the
inauguration of a true philosophy has revealed their

glaring inconsistencies. At this very day we read in the newspapers often of some comet which is certainly about to destroy the present state of things; while, even in case no such apprehension is announced, we are charmed with the cheerful intelligence that in the German vineyards, in the beautiful valleys of the Rhine and Moselle, these once ill-omened bodies have been exercising a beneficial influence on the ripening of the vine.

It is indeed pleasing and instructive to consider the important facts which history has furnished us on this subject; but the time has now come when all such fears as we have noticed should be at an end, and when the return of a comet, which formerly spread dismay and terror throughout the world, should be regarded as one of the greatest triumphs of science.

> "Lo! from the dread immensity of space
> Returning with accelerated course,
> The rushing comet to the sun descends;
> And as he sinks below the shading earth,
> With awful train projected o'er the heavens,
> The guilty nations tremble. The enlighten'd few
> Whose godlike minds philosophy exalts,
> The glorious stranger hail. They feel a joy
> Divinely great; they in their powers exult,
> That wondrous force of thought, which mounting spurns
> This dusky spot, and measures all the sky;
> While from his far excursion through the wilds
> Of barren ether, faithful to his time,
> They see the blazing wonder rise anew."

CHAPTER II.

BEFORE proceeding to give a detailed account of
some of the most remarkable comets which have
appeared during the period which has elapsed since
the earliest records of the Chinese, it may not be
improper to explain, in this connection, some of the
general and more prominent characteristics by which
these bodies are distinguished. We shall subsequently enter into a complete exposition of the
theory of their form and physical constitution, and
we propose to notice now simply what may be
required to render the descriptions which follow
intelligible.

The comets, for the most part, present the appearance of a globular mass of illuminated vapor or
nebulous matter, with a train, or *brush*, as it is
termed by the Chinese astronomers, extending in a
direction opposite to the sun. There are, however,
those which, being invisible to the naked eye, exhibit no signs of a train or elongation opposite the

7 *

sun, and of which even the nebulous mass itself is ill-defined and irregular. The latter are generally of extreme faintness, and evince their cometary nature only by their motion in their orbits. The discovery of these has been much more rapid of late years than formerly, owing to the perfection and number of the instruments employed in searching for them; and although, as will hereafter be noticed, these apparently insignificant objects are by no means uninteresting, yet it will be sufficient for our present purpose to understand the phenomena presented in the case of the larger and brighter telescopic comets, and those visible to the naked eye.

The nebulous mass of matter, independent of the train, is called the *head* of the comet, while the latter is usually termed the *tail*. At or near the centre of the head there is almost always a point of greater condensation of light than in the other parts of the comet; and this bright spot, which is not invariable, is called the *nucleus*. The nucleus appears, in the case of some of the great comets, sometimes like a minute stellar point, and sometimes presents the appearance of a planetary disc shining through a nebulous haze, which is termed the *coma*.

These definitions will serve to illustrate the descriptions which we shall now give of some of the

most remarkable comets whose appearances have
been recorded; and, as we have already remarked,
we shall subsequently explain, in the proper connection, all the various changes which take place in
the figure and constitution of the nucleus and train
as the comet approaches the sun, and in receding
from it.

The Chinese have recorded the appearances of
comets professedly as far back as the year 2550
before the commencement of our era; but the
accounts given are so vague and unsatisfactory, and
the dates so uncertain, that although these bodies
have not failed to attract a great deal of attention
even in the very remotest ages of the world, yet we
have no authentic observations earlier than about
four hundred years before Christ. There can be no
doubt, however, but that the Chinese have actual
records far anterior to this date; but so little is
known in regard to them, that we can glean nothing
from the accounts given which could interest a
general reader.

During the four centuries which immediately preceded the birth of Christ, we have reliable accounts
of the appearances of thirteen comets. Some of
these are described as having been of prodigious
magnitudes, and as having presented terrific aspects.
So little was known in those days in regard to their
nature and motions, that the phenomena recorded

give us no additional information in regard to them which is of any value. As usual, they were supposed to have caused the most dreadful disasters which occurred at or near the date of their appearance; and if the accounts are not exaggerated, the phenomena presented were fully as imposing as any which have been exhibited in modern times. It is related by Diodorus, that at the time of the expedition of Timoleus of Corinth to Sicily, in the year 344 before our era, a burning torch was visible in the heavens throughout the entire night, and that it preceded the train of Timoleus until its arrival in Sicily, thus connecting indirectly the appearance of the comet with this particular event. Seneca records the appearance of a comet in the year 156, which is said to have been apparently as large as the sun. At first its appearance was red and fiery, emitting a sufficient quantity of light to dissipate the darkness of night; but it afterwards diminished gradually, and finally became invisible.

The year 136 B. C. was signalized by the appearance of three large comets, one of which was so bright that the heavens appeared to be on fire, the comet being described as having occupied one-fourth part of the sky. Lucian informs us that a comet, which appeared in the year 48 B. C., was so brilliant that the darkness of the night was illuminated by it, and that the comet itself presented a terrific appearance. It is also recorded that a comet appeared in the year 43 B. C., which was visible two or three hours before sunset. Dion Cassius relates that, under the consulate of M. Valerius Messala Barbatus and P. Sulpicius Quirinus, — which corresponds to the year 11 B. C.,— before the death of Agrippa, a comet was seen for many days. It appeared as if suspended over the city of Rome, and afterwards resolved itself into many smaller comets. The latter statement might cause us to feel inclined to reject the account given by Dion Cassius, were it not that we have already an example of a similar character in the case of Biela's comet, which will be described hereafter.

The foregoing descriptions relate to the principal comets which appeared anterior to the commencement of our era. During the five centuries immediately following this epoch, we have records of the appearances of no less than eighty-three comets, some of which are described as having been truly magnificent. A very beautiful one appeared in China, in the year 178. It was first seen on the 30th of August, and exhibited a tail only a few degrees in length. It became, however, gradually brighter, until the train extended over more than sixty degrees of the heavens. The nucleus was remarkable for its ruddy appearance, and remained visible nearly three months; having, in the meantime, passed over an arc of three hundred degrees in its apparent course.

In the year 389, a comet of great brilliancy appeared in the north-east, a few hours before sunrise. It is described as resembling a lamp, with the flame tending upwards towards the zenith. It was visible during a period of forty days, and finally disappeared in the constellation Ursa Major. A comet of precisely the same aspect, and moving in the same direction, was visible in the following year. Claudian describes a comet which was visible in the year 402, as being first seen in the east, towards that part of the heavens in which the constellations Cepheus and Cassiopeia are situated. It subsequently moved northward above Ursa Major, and was so brilliant that it diminished the beauty of the bright stars of that constellation. It then began to fade away, and finally vanished in a very narrow flame.

On the 19th of July, 418, there was a total eclipse of the sun, during which the darkness was so great that the stars were plainly visible. During the time of total obscuration, a light in the form of a cone was perceived in the sky, which resembled the flame of a lamp. This object proved to be a comet which was subsequently seen in China, where it presented a very beautiful appearance. When it first became visible in China, it was in the constellation Cygnus, and moving west and north through the Great Bear:

it finally disappeared in the constellation *Leo*, having been visible more than three months.

After the commencement of the sixth century, the recorded appearances of comets are much more frequent; and we shall attempt to notice only those which were most remarkable, either on account of their extreme brilliancy, or from some peculiar physical characteristic. It should be remarked, however, that until about the year 1200, we are compelled to rely almost entirely on the Chinese accounts, which are often, if not generally, vague and unsatisfactory. The great social and political convulsions which were continually disturbing the nations of Europe, were hostile to the advancement of astronomical science; and while the most enlightened nations of the globe were absorbed wholly in the study of eloquence, poetry, and the fine arts, and more especially in the pursuit of military fame, it was reserved for the inhabitants of the East to cherish science in its infancy, and to furnish the elements of its progress. When the affairs of Europe had become settled, the study of astronomy soon began to engage the attention of learned men; and henceforward we are provided with the most complete and accurate descriptions of the comets which appeared.

In the year 530, a very large and brilliant comet appeared in China, which was found to move rapidly toward the north and west. It was first seen in October; and, since a comet was seen in Europe in the beginning of the year 531, presenting the same general appearance, we may conclude that the comet was visible in China before its passage through the point of its orbit nearest the sun, and in Europe after its passage through that point. This comet has been supposed to be identical with that which appeared at the death of Julius Cæsar, and which Whiston supposed to have caused the Mosaic deluge, as we have already noticed. It is related by Gregory of Tours, that in the year 582, on Easter day, at Soissons, the heavens were seen on fire, — which remark is supposed, in all probability, to refer to a very large and terrific comet, which first made its appearance in January of that year. It was surrounded by an immense, nebulous envelope, such that the comet is said to have been situated, as it were, in a kind of opening. Its train extended to an enormous distance from the nucleus, and is described as having resembled the smoke of a great conflagration, viewed from a distance. The comet was seen in the west immediately after sunset.

On the 8th of July, 615, a comet was seen in China, in the Great Bear. Its tail was fifty or sixty degrees in length, and was remarkable for its dusky color. During the night its head or nucleus appeared to have a motion of libration. It moved to the north-west for several days, and afterwards retrograded and disappeared. A comet is mentioned by European historians about the time of the election of Pope Donus, in the year 676, which had a tail five degrees in length, and was remarkable from the fact that its motion was such that its gradual diminution from day to day, as the comet receded from the earth, was distinctly noticed. This comet was also observed in China, on the 4th of September, 676, in the constellation *Gemini*. It moved toward the north-east, and disappeared about the 1st of November in the constellation *Ursa Major*.

Father de Maille relates, that on the 22d of March, 837, a comet was seen in China in the adjacent part of the constellations *Aquarius* and *Pegasus*, the tail of which was seven degrees in length. It was found to move toward the west, and on the 6th of April the length of the tail was ten degrees. On the 10th of April the length of the tail had increased to fifty degrees, and was separated into two distinct parts, the bifurcation commencing at the nucleus and increasing in width to the end of the tail. On the 12th of April, its length was sixty degrees, and appeared again undivided, the secondary train, which was not as bright as the primary, having apparently vanished. On the 14th of April, the tail extended from the horizon to the zenith, and the appearance of the comet was extremely grand. After this date

8

the length of the tail decreased rapidly, and on the 28th of April the comet was seen for the last time, with a train only three degrees long. Some astronomers have supposed this comet to have been a former appearance of a comet which appeared in 1682, and which is now known as Halley's comet.

One of the largest and most remarkable comets on record made its appearance in the year 895. It was seen in China on the 25th of June, in that year, and the tail or brush is said to have been one hundred degrees in length. When the comet first became visible it was in the Great Bear, but it subsequently moved towards the constellation Hercules. The brilliancy increased very rapidly, and, according to the Chinese historians, it actually attained the enormous and unexampled length of two hundred degrees. This account is undoubtedly somewhat exaggerated, but it is sufficiently certain that the length of the train was unsurpassed by any other which had ever before been recorded. Another large comet appeared in the autumn of the year 975. In China it was seen on the 3d of August, in the constellation Hydra, at which time the tail was forty degrees in length. The comet was visible during a period of eighty-three days, and passed through the constellations Cancer, Gemini, Taurus, and Aries, and a part of Pegasus and Andromeda. The course pursued by the comet was such that

Pingré supposes it to be identical with the great comet which appeared in 1264 and in 1556. By supposing that the comet passed nearest the sun a few days before the end of July, 975, it may be shown that it would have been in conjunction with the sun on or about the 3d of August, but that its north latitude being considerable, it would rise before the sun, and might exhibit all the phenomena which have been recorded.

In April, 1066, a comet of great brilliancy was seen both in Europe and in China. It was supposed by European historians to be the forerunner of the conquest of England, by William, Duke of Normandy. The course of the comet seems to render it probable that it is a previous return of the comet of 1677. Another comet was seen for a very short time, in October, 1097. It was visible in Europe only fifteen days, and is remarkable on account of having two separate trains, which were directed toward the east and south-east respectively. In China, on October 6th and 9th, its tail was noted as being respectively thirty and fifty degrees in length. On the 6th of October it was seen in the constellation Libra, and on the 16th of the same month it had moved north as far as the head of Hercules. It ceased to be visible in China on the 25th of October. A large and brilliant comet was also seen in Palestine and in China early in February, 1106. Its tail was sixty degrees in length, and was similar in color to the whiteness of snow. The comet was visible during a period of fifty days.

In the year 1264, a great and celebrated comet made its appearance, and its magnitude was such that it is mentioned by all the historians of that day; and it is also distinctly stated that no one living at the time of its appearance had ever seen one which could be compared with it. It was first seen about the 1st of July, and attained its maximum brilliancy in the latter part of August and in the beginning of September. The tail was one hundred degrees in length, and appeared curved in the form of a sabre. The train was visible in the east early in the evening, although the nucleus did not appear above the eastern horizon until near morning. The comet was last seen on the evening of the 3d of October, the date of the death of Pope Urban IV., of which event it was considered the precursor. The course pursued by the comet was such that, in more modern times, it has been conjectured that this body is identical with the great comets which appeared in 975 and in 1556.

We have here, and also in the preceding pages, alluded to the supposed identity of comets which have appeared at different epochs. These allusions may seem premature, from the fact that, thus far, no account has been given of the nature and forms of the paths pursued by the comets in their motions through space. For reasons which will presently be made obvious, it has seemed expedient to defer a complete exposition of the theory of the motions of the comets, until we come to treat of those comets which are definitely known to have fixed periods of revolution, and which have been certainly identified at two or more different visits to our vicinity. Under such circumstances it may be advisable to state, at once, in reference to the forms of the cometary orbits, what will suffice for our present purpose, reserving for the proper connection a detailed explanation of the theory of the motions of these chaotic worlds.

The comets are known to move in regular orbits in three different kinds of curves, in obedience to the law of universal gravitation. These curves are the ellipse, parabola, and hyperbola, the nature of which will be subsequently explained. Those which move in ellipses will return regularly to the sun at stated intervals, while those which move in parabolas and hyperbolas visit our system but once, and thence pass on to visit other suns and systems. Of all the comets which have been discovered, a majority are found to move in these curves, and the number which move in elliptic orbits whose dimensions are confined within the probable bounds of the solar system, is much more limited. But a small

number, comparatively, have been observed at more
than one appearance, and these we propose to consider as a separate and distinct class. There are
many others which the delicate refinements of
modern astronomy have shown to move in orbits
whose ellipticity is clearly indicated, but in which
the deviation from either the hyperbola or parabola
is so uncertain that it is impossible to predict,
within very narrow limits, the precise date of their
return. The comet of 1264 is one of these; and
although the elements which determine the form and
position of its orbit in space, are found to bear a close
resemblance to those which have been obtained for
the comets of 975 and 1556, yet we are unable to
venture a prediction that these comets are identical,
and that another appearance may be expected after
the lapse of about 300 years from the date of its last
visit to the sun. The point of the orbit which is
nearest the sun is called the perihelion; the point
which is most remote is called the aphelion; and
the nature of these curves is such, that for a considerable distance on either side of the perihelion
they nearly coincide. This, together with the uncertainty of the observations, and the fact that the
comet is visible from the earth only when near the
sun, sometimes renders it impossible to decide in
regard to the exact form of the orbit. With these
explanatory remarks, we will continue the descrip-

tion of the most remarkable comets which have appeared subsequent to the one we have just noticed.

In the year 1402 a very large and brilliant comet made its appearance. It was seen at different places on the 8th of February, and was so brilliant that it could be seen at mid-day. It disappeared early in March. Another comet of extraordinary magnitude made its appearance in 1456. It is represented by historians as being grand and terrible, with a tail more than sixty degrees in length. It was visible during the month of June, and was very remarkable on account of the rapid variations in the size of the train, which at one time did not exceed seven degrees in length. The first appearance of the comet was so sudden and terrific, that a belief became very prevalent, among all classes, that it would destroy the earth, and that the Day of Judgment was at hand. This is the comet which we have already noticed as having created a most profound sensation at Rome, since it was regarded as having presaged the rapid success of the Turks, who were then engaged in subjugating the nations of Europe. The fact that the followers of Mohammed had already crossed the Hellespont, and that no force had been able to check their progress, together with the great apprehension which the appearance of the comet had already excited, seemed to add to the general gloom which pervaded all classes of society; and it was under these circumstances that Pope Calixtus II. ordered prayers to be offered daily for preservation from the baneful influences of the comet, the devil, and the Turks, and issued a bull in which all three were equally anathematized. This comet is now recognized as a previous return of the comet of 1682, known as Halley's comet; and its great brilliancy resulted from the favorable circumstances, in this respect, of its position with reference to the earth and sun.

In February, 1490, a comet appeared in Europe which had a small nucleus, but a train extending to an enormous distance, and remarkable for its phosphorescent appearance. It was visible at Boulogne about the middle of February, and was probably not seen in China and Japan. A comet is recorded in the Chinese annals in the year 1491, but there is hardly any reason to believe it to have been identical with this.

The comet of 1456 again made its appearance in July and August, 1531, but was far from the earth and sun, and consequently not very brilliant. The year following, 1532, a comet was seen in China and also in Europe, which has been supposed to be identical with one which appeared in 1661. It was visible in the morning before sunrise, and the nucleus is described as being apparently three times larger than Jupiter. It was visible in Europe about seventy days, and in China one hundred and fifteen days. The train varied in length from one to ten degrees, and was very bright.

In 1556 another comet of prodigious magnitude made its appearance. It began to be generally visible about the end of February, at which time it is said to have been more than half as large as the full moon. Its tail, however, was short and variable, and for this reason it was described as exhibiting a movement like that of a flame, or a torch disturbed by the wind. The greatest length of the tail did not exceed four degrees, and its color was ruddy, not unlike that of the planet Mars. It moved rapidly north, and on the 12th of March was distant from the earth only eight millions of miles. The comet disappeared in the sun's rays on the 23d of April, in the southern part of the constellation Cassiopeia. It created considerable excitement in Europe, and is said to have been the immediate cause of the abdication of Charles V., Emperor of Germany.

The orbit of this comet was computed by Dr. Halley; but since the observations on which they were based were uncertain, they have not been considered very accurate. They were sufficient, however, to indicate the probable identity of this comet with those which had been observed in 975 and

1264. He was therefore led to conjecture that the comet had a period of revolution around the sun of about 292 years, and that it might be expected to reappear in 1848. A few years later Pingré collected together all the observations of the comets of 975 and 1264, and more especially the latter, which could be found in both the European and Chinese annals, and from a complete investigation of the elements of the orbits of the comets of 1264 and 1556, he found that the courses of both comets, at the date of their appearance, could be very satisfactorily represented by the same elements. He therefore confirmed the hypothesis of Halley, and concurred in the opinion that the comet might be expected again in 1848. Nothing further was attempted until the time had nearly elapsed at which it was to reappear. Between the years 1843 and 1847 the whole subject was investigated anew by Mr. Hind, of London, and M. Bomme, of Middleburg, in the Netherlands. Mr. Hind found that the elements were such as to confirm their supposed identity; and by taking into account the planetary perturbations, he concluded that the comet would appear about the year 1858.

The computations of Bomme were more extended, and he found that with Halley's elements, at the time the comet was visible in 1264, it was moving in an ellipse, with a periodic time of 112,469 days,

or about 308 years; but that the disturbances produced by the attraction of the planets would accelerate its motion, and shorten its period 5903 days, so
that it passed its perihelion in April, 1556. He
found that at this return it was describing an elliptic
orbit which gave a period of 112,943 days; but that
the next revolution would occupy a period of 111,146
days, bringing the comet to its perihelion on the
22d of August, 1860.

With Hind's elements it was found that, in 1264,
the ellipse described by the comet corresponded to
a period of 110,644 days, or nearly 303 years; and
that the attraction of the planets would shorten its
period a little more than eleven years, thus bringing
the comet to its perihelion in 1556. The same elements indicated that at the time it was visible in
1556, its mean motion corresponded to a period of
308 years, and that the period of its next return
would be expedited by planetary disturbances 3828
days, or ten and a half years. The comet would,
therefore, arrive at its perihelion early in August,
1858. Thus it appears that there is an uncertainty
of at least two years in the time of its next appearance; and since the interval included between the
results of Hind and Bomme has now elapsed, we
must conclude that those of the latter are most accurate, although themselves perhaps very far from the
truth.

The important circumstance of an apparent identity between the comet of 975 and those of 1264 and 1556, has already been noticed; and it is found that, assuming the identity, and adopting Hind's elements, the path of the former may be very closely represented. This question, however, has been elaborately discussed by M. Hoek, of Leiden, and he concludes that there is no reason to regard the comets as identical. He finds that the path of the comet cannot be satisfactorily represented by the elements of either the comet of 1264 or 1556, each of which he has investigated anew. He finds also that there exist great discrepancies even between the orbits of the latter comets, and announces that their identity is extremely doubtful. It appears, therefore, that it is barely possible that the comet of 1556 may shortly appear, although no such prediction can be entitled to any particular confidence. Should the comet really make its appearance, its future movements would no longer be involved in doubt, but its next and succeeding returns could be predicted with the utmost precision.

The next comet of any considerable magnitude, compared with those which we have already described, made its appearance in 1590. It was discovered by Tycho Brahe, on the 5th of March, and observed until the 16th of the same month. It is noted as being of a medium magnitude, but had a great tail, which extended as far as the zenith, the nucleus being near the horizon. On the day of its discovery, the comet appeared as bright as a star of the second magnitude; and shortly afterwards, on the same night, it was equal in brilliancy to a star of the first magnitude. Another large comet appeared in 1618. It was discovered in Silesia on the 10th of November, and at Rome on the same evening. It was also observed by the Spanish Ambassador at Ispahan, in Persia, for fifteen days, commencing with the date of its discovery in Europe; and presented a magnificent appearance in the eastern sky, about two hours before sunrise. The length of the tail was about 60 degrees.

In December, 1652, a comet was seen in Europe, which was of a pale and livid color, and almost equalled the moon in magnitude and brilliancy. No mention is made of the length of its tail. Another comet of considerable size was seen in Spain, on the 17th of November, 1664. It was subsequently observed in France and Germany; and was last seen by Helvetius, at Dantzic, on the 18th of February, 1666. At its first appearance, it was noted as large as a star of the first magnitude, but not so bright. The length of the tail varied from five to ten degrees. Four years later, 1668, the tail of a comet was seen above the southern horizon, at different places in Italy. The nucleus was not visible, but the comet was observed in the southern hemisphere. This comet has been supposed to be identical with the great comet which appeared in 1843, and is remarkable for its near approach to the sun.

One of the most remarkable comets on record made its appearance toward the close of the year 1680, and was carefully observed by the astronomers of Europe. It is said to have appeared to descend, as it were, from the distant regions of space with a prodigious velocity, almost perpendicular to the sun, and to have ascended again in the same manner, with a velocity retarded as it had before been accelerated. The comet was discovered at Coburg, in Saxony, on the 4th of November, 1680; but the name of the discoverer is unknown. It was also discovered independently by Godfrey Kirch, on the 14th of November, while about to observe the moon and Mars. It was very carefully observed at Paris by Cassini, during the entire period of its visibility. It was observed by Flamstead, at Greenwich, until the 15th of February, 1681. The observations of Cassini, Flamstead, and others, enabled Newton to trace out the form of its orbit, and its position in space; and he found that the observations could be satisfactorily represented by supposing the orbit to be a parabola, with the sun in the focus. At its perihelion, he found that the distance of the comet from the surface of the sun did not exceed one-sixth of the sun's diameter, or about 147,000 miles. The velocity of the comet, when nearest the sun, must therefore have been 880,000 miles an hour.

A more complete determination of the elements of the orbit of this comet was made by Dr. Halley, who found that the orbit was sensibly an ellipse, with the sun in one of the foci; and that the perihelion distance was only 590,000 miles from the sun's centre, while its aphelion distance was at least 13,000,000,000, or thirteen thousand millions of miles. He determined also that, in passing the point of intersection of its orbit with the plane of the earth's orbit, in moving from the north to the south side of the latter, or in passing through what is called the descending node, the comet was distant from the earth only 440,000 miles. It passed through this point on the 22d of November; and it was remarked by Halley that, had the earth been then in that part of its orbit nearest the descending node of the comet, their mutual gravitation would have caused a change in the plane of the earth's orbit, and in the length of our year; and if so large a body, with so rapid a motion, were to strike the earth, the shock might reduce this beautiful structure to its original chaos.

Whiston supposed this comet to be identical with the comets which appeared in the years 1106, 531, and 43, B. C.; and under this supposition he com-

putes backward, and attempts to explain the phenomena of the Mosaic deluge by a near approach
of the comet to the earth, as already explained.
The computations of Euler, Pingré, and Encke, have
fully confirmed the supposed ellipticity of its orbit, but
indicate that its period of revolution must be vastly
greater than 575 years; and that, consequently, the
speculations of Whiston, at least so far as this comet
is concerned, must be regarded as extremely absurd.
This comet was remarkable for its near approach to
the earth and sun, and also for the brilliancy of its
light and the length of its tail, which extended over
half the vault of heaven, or from the zenith to the
horizon.

In the month of August, 1686, a comet was visible
at Para, in Brazil, whose nucleus was as bright as a
star of the first magnitude. The tail was eighteen
degrees in length, and is described as having presented a very beautiful appearance. Three years
later a comet of uncommon splendor was seen in
China and at different places in the southern hemisphere. It was first discovered at Pekin on the 11th
of December, 1689, but was not visible in Europe.
On the evening of its discovery the nucleus was very
brilliant, and the tail was about twelve degrees in
length. In the southern hemisphere the comet was
much more favorably situated for observation, and
consequently its general appearance was correspond-

GREAT COMET OF 1744

ingly improved. It was equal in brilliancy to the
bright stars of the first magnitude, and the greatest
length of the tail was 60 degrees.   Another comet,
similar in size and appearance to that of 1686, just
described, was seen in Brazil on the 28th of October,
1695.   The length of the tail was 18 degrees, and
the nucleus was remarked to be at first almost obscured by the atmosphere or coma by which it was
surrounded, and was even, at intervals, scarcely distinguishable.

The first part of the eighteenth century was not
remarkable either for the number or brilliancy of
the comets which appeared.   There were, however,
several of sufficient size and brilliancy to be visible
to the naked eye, but not such as to present a very
striking appearance.   It was not, therefore, till 1744
that a comet was visible which deserves a notice in
this connection.   This comet was discovered at
Harlem on the 9th of December, 1743, and was observed at several observatories, until about the time
of its perihelion passage, which took place on the
evening of the 1st of March, 1744.   It was one of
the largest and most beautiful comets which had
appeared since the famous one of 1680, and therefore excited no small degree of interest throughout
the civilized world.   The nucleus was very bright,
and although the tail was not remarkable for its
length, yet it was noticed by Cassini, early in Feb-

9 *

ruary, that the head was separated into two distinct
parts, and the train exhibited decided symptoms of
a corresponding appearance. The tail was, indeed,
shortly after this observation of Cassini, observed
to be divided into two branches. At one time subsequently, and just before its passage through the
perihelion of its orbit, it is said to have had six
separate and distinct tails, the lower one relatively
to the horizon being greatly curved near the extremity. The accompanying diagram represents the
appearance of the comet at this time. The remarkable phenomenon, however, of six distinct tails was
of only a few days' duration, and the comet appeared
again as before, having but two trains. This comet
approached the sun to within twenty millions of
miles, and the elements of its orbit which have been
computed do not seem to indicate that it will ever
reappear to human vision.

The year 1748 was remarkable for the appearance
of two bright comets at nearly the same time. The
first was discovered at Paris about the end of April
of that year, and was also observed in South America, and in China. It is described as having been
a fine object, easily seen with the naked eye, and
having a tail twenty degrees in length. It was last
observed by Maraldi, at Paris, on the 30th of June.
The second was discovered at Harlem, on the 19th
of May, in a different part of the heavens from that

in which the one just described was seen. Its nucleus
was brighter than the preceding, but there was no
appearance of a tail. From this date until 1769, if
we except a reappearance of the comet of 1682,
known as Halley's comet, which will be described
hereafter, no comet of considerable size was observed.

The great comet of 1769 was discovered by Messier, at Paris, on the 8th of August, and was subsequently observed throughout Europe. It was observed by La Nux at the Isle of Bourbon, in the
Indian Ocean, from the 26th of August to the 26th
of September; and all the various changes which
took place in the form and size of both the nucleus
and the tail were carefully recorded. On the 11th
of September the length of the tail was found to be
97 degrees, while on the 28th of August it did not
much exceed 15 degrees. On the 9th of September
the length of the tail was variously estimated from
43 to 75 degrees. It passed nearest the sun on the
7th of October, about midnight, at a distance of
eleven millions of miles. It was nearest the earth
about the middle of September, and was distant
about ten millions of miles. It was last seen on the
1st of December, having been visible more than
three months. At the time of the appearance of this
comet, the methods of determining the form and
position of the orbits of these bodies in space by

means of observations made during a few days, had
been successfully investigated; and since the astronomers were enabled to predict from day to day, and
even for several weeks in advance, the variety and
grandeur of the phenomena which it would present,
even to the unassisted eye, it very naturally excited
a lively interest among all classes of society. The
tail was not actually as long as had been exhibited
in the case of many other comets which had appeared; but the near approach of the comet to both
the earth and sun, and at very nearly the same time,
gave it a most magnificent appearance, and caused
it to appear even greater than it really was. The
researches of modern astronomers seem to indicate
that the comet will return again to our system after
the lapse of a few thousand years, but fail to indicate
within any very close limits the exact period of its
return.

During the last half of the eighteenth century,
no less than forty-five comets were observed; but,
with the exception of Halley's comet, and that just
described, they were not remarkable for their brilliancy. Some of them were bright enough to be
distinguished by the naked eye, and sometimes exhibited a faint train a few degrees in length; but
the great majority were round and ill-defined nebulosities, with a slight condensation of light at the
centre.

The next comet of considerable size was discovered by Pons, at Marseilles, on the 20th of September, 1807. This is the date of the first authentic
observation, although it seems to have been seen
eight days previously by an Augustine monk in
Italy. This was the finest comet which had appeared since that of 1769. On the 30th of September
the nucleus was nearly as bright as a star of the
second magnitude, and the tail was distinctly visible.
On the 8th of October it was seen in the north-west,
soon after sunset, and not far distant from Arcturus,
which was then only a few degrees above the horizon. It is described as presenting to the naked eye
the appearance of a dim nebulous star of the second
magnitude, with a beam of light on one side of it.
Through a telescope its tail presented a brilliant
appearance, and was more than a degree in length.
The coma, and even the nucleus, were remarkably
ill-defined. On the 7th of November the tail was
seen divided into two parts; and, a few days later,
the comet ceased to be visible to the naked eye. It
was last observed on the 27th of March, 1808, at
which date it was a very faint telescopic object.

The orbit of this comet has been very carefully
determined by Bessel, and he finds that it is an
ellipse, corresponding to a period of revolution of
about 1500 years. The comet passed the perihelion
of its orbit in the forenoon of the 19th of Septem-

ber, at a distance from the sun of about 62,000,000
miles. The aphelion distance is about eleven thousand millions of miles. The diameter of the nucleus
was about 4600 miles, or about the size of the planet
Mars, and seemed to be of considerable density.
The diameter of the coma or envelope was estimated
at 120,000 miles, but appeared to vary at different
dates; and the velocity of the comet in its orbit,
when nearest the sun, was considerably more than
two millions of miles daily. The tail, as already remarked, was divided into two separate branches.
The north side, however, was much brighter and
better defined than the other, and was also invariably
convex, while the other was concave. The most
remarkable feature about the tail was the rapid
variations which took place in its length and brilliancy. It was seen at times to exhibit coruscations
or flashes of light, not unlike what is exhibited in
the case of the aurora borealis. It is asserted that
in less than one second of time, streamers might be
seen to shoot out from the body of the tail, and
generally from the vicinity of the nucleus, to a distance of two and a half degrees, and that they as
rapidly disappeared and issued out again. These
coruscations may perhaps have been much less
frequent than what is here stated, since the accounts
given are very probably exaggerated; but, as will
be evident in the case of comets which appeared

subsequently, there can be no doubt but that streamers were observed to reach out even as far as the whole length of the tail, or to a distance of nearly five millions of miles. The light of these streamers was also described as being sometimes' whiter and clearer at the end than at the base.

We come now to consider the case of the great comet of 1811, which was discovered by Flaugergues, at Vivières, on the 26th of March of that year. When first discovered it was a faint nebulous object, with a slow motion as seen from the earth, and far distant from both the earth and sun. It did not exhibit any signs of a tail, but presented simply a slight condensation of light at the centre. It continued visible until the 10th of June, when it was lost in the approaching twilight. It had, however, been already extensively observed by the astronomers of Europe; and from the observations it was found, by computing its orbit, that it would not pass its perihelion until about the 12th of September, and that it would again emerge from the sun's rays early in August. It was found, also, that it was rapidly approaching both the earth and sun, and that it would certainly present a magnificent appearance during the entire months of September and October. These predictions were indeed realized to the very fullest extent. The comet was again detected on the 20th of August, and observed almost without interruption till the 11th of January, 1812, when it was again lost in the twilight. It was, however, detected the third time by Wisniewsky, in Neu-Tscherkask, on the 31st of July, 1812, and observed for the last time on the 17th of August of the same year, having been visible for a period of nearly seventeen months. It was visible to the naked eye during a period of more than three months; and, on account of its immense size and unexampled brilliancy, excited a most profound sensation. To astronomers it was especially interesting, on account of the facilities which its long visibility afforded for the determination of the elements of its orbit, with the very greatest precision; and, also, for the opportunity of observing its aspect and its physical constitution. The great perfection to which theoretical and practical astronomy had been carried at the date of its appearance, and the great number of observers, provided with superior instruments, rendered it possible to investigate with great facility all the various questions which might arise. Accordingly, we find by reference to scientific periodicals, that all the varied phenomena presented by the comet have been carefully recorded, as well as accurate determinations, from night to night, of its position in the heavens.

The comet passed the perihelion of its orbit, or the point nearest the sun, on the 12th of September,

at a distance of 98,565,000 miles. On the 7th of
September the tail appeared to be bent off in two
branches. These branches, however, did not proceed from the head of the comet, but seemed to be,
as it were, hung together at a slight distance from
it, and separated from it by a dark interval. The
tail was then five degrees in length, but on September 20th it had increased to 10 degrees. On the 11th
of October the length of the train was only about
13 degrees, while a few days previous it was found
to be 25 degrees long, and 6 degrees broad at the
extremity. The central condensation of nebulous
matter was 50,000 miles in diameter, or more than
six times the diameter of the earth. The envelope
was 30,000 miles in thickness, and the centre of the
nucleus was separated from its interior surface by a
space of 36,000 miles, so that the diameter of the
head of the comet must have been 132,000 miles.
In the centre of the nucleus there was a brilliant
point 428 miles in diameter, which was probably
composed of solid matter. The part of the nucleus
immediately surrounding this may be supposed to
have been fluid, and those portions still more remote
to have been composed of a sort of gaseous or ele-
mentary form of sidereal matter. It was imagined
by some that the comet had a sort of phosphorescent
light of its own, proceeding from the denser portion
of the nucleus. The coma was extremely rarefied,

and presented the appearance of a very faint whitish light, scattered in separate portions. At the time when the train was seen divided into two parts, this apparent division of the coma was very perceptible, and the part which directly encompassed the nucleus was much brighter than the other portion, which shone with a faint greyish light, and sweeping around the former at a distance, formed the double tail. The phenomena thus presented made the comet appear somewhat like a very brilliant object surrounded by a dense net-work of gauze. The diameter of this exterior envelope was more than 500,000 miles, and its appearance was such as to induce Schroeter to suppose that it very clearly indicated the existence of a repulsive force residing in or around the nucleus, but of what character he was unable to determine. Toward the latter part of November and in the first week in December, great changes were observed to take place in the form and magnitude of the head of the comet. The nebulous matter, which had for several months exhibited the appearance of extreme rarefaction, was seen to be condensed or attracted towards the nucleus; and the changes were so rapid that there was exhibited the most incontrovertible proof of physical action on a grand scale, perhaps such that no phenomena which have ever been witnessed on our earth, will serve to give us any adequate conception of its magnitude.

The position of the comet, when brightest, with respect to the earth and sun, was such that although the train at one time extended to a distance of 120,000,000 miles, it did not subtend an angle of more than 20 degrees. Under more favorable circumstances as regards position, it might have been 50 degrees in length. In the case of this comet, also, streamers were seen to shoot out from the convex side of the tail to a distance of more than two millions of miles, thus producing to a limited extent the appearance of coruscations such as were presented by the comet of 1807. The breadth of the tail, at its extremity, was nearly fifteen millions of miles. The comet approached the earth only to within about 110,000,000 miles. It was supposed by Herschel, that the solid portion of the nucleus was spherical, that it shone in part by its own native light, and that it probably had a rotation around an axis.

The elements of the orbit of this comet have been investigated by several astronomers, and it has been found to be elliptical, although the periods of revolution assigned by different computers vary between long limits. The most complete research, however, was performed by Argelander, who found, from a discussion of the entire series of observations worthy of confidence, that the period of revolution of the comet is 3,065 years. Bessel, by a similar process of investigation, and taking into account the disturbances produced by the attractions of the planets — which will subsequently be explained — finds its period to be 3,383 years. Its aphelion distance is, therefore, more than one hundred and fifty thousand millions of miles. But great as this distance is, it is perfectly certain that there are many comets which revolve in orbits far more extensive than the one described by this comet. Indeed, there seems to be no limit to the distance to which these erratic worlds may sweep outward from the sun; and their return depends simply on the fact, whether they recede so far as to fall within the attractive influence of some other sun, toward which they begin to urge their flight, and through whose system of planets they carry the same apprehensions of danger which they have caused in our own system, or more especially to the inhabitants of our earth. Thus may the comets be regarded a connecting link between the systems of suns and planets which fill the immensity of space.

The great comet of 1811 may justly be considered as the most magnificent one, in all respects, which has ever visited our sun. In the case of nearly all the very bright comets, the perihelion distance has been small; but in the case here presented it was considerably more than the mean distance of the earth from the sun. The other bright comets have usually approached very near the earth, while this one was even more remote from us than from the sun. Yet under all these circumstances its brilliancy has rarely, if ever, been excelled, while in magnitude it was surpassed only by the sun itself. It is no wonder, therefore, that to those unacquainted with the nature of such bodies, it was a source of great alarm; or that even among the more intelligent population of Europe, it was regarded as having produced a most wonderful effect on the climate of the earth; although astronomers have shown that such was not the case.

For a period of thirty years immediately following the disappearance of the comet of 1811, the discovery of these mysterious bodies was of frequent occurrence; but only in a few instances were they of sufficient brilliancy to be visible to the naked eye. Early in July, 1819, there was one discovered in Europe which became pretty bright, and had a tail about 8 degrees in length. In the years 1821, 1822, and 1823 there were comets visible to the naked eye, one in each year, and the last one was remarkable for having exhibited the phenomenon of having, in addition to the usual tail, another one in a contrary direction, varying from four to seven degrees in length. In 1825 and 1826 there were comets of considerable size; and in 1830 there were two which were distinctly visible to the naked eye, one of

10 *   H

which, discovered at sea on March 17, had a tail
about 8 degrees in length, with a very bright nucleus.
In 1835 the comet of Halley made its appearance, in
accordance with the prediction; and, in the southern
hemisphere, presented a very fine appearance.   On
the 6th of March, 1840, a comet was discovered by
Galle, at Berlin, which had a tail several degrees in
length.   These comprise the principal comets, with
the exception of two or three periodic ones to be
described hereafter—which had appeared from 1811
up to this date, and do not afford any instance of
one of sufficient interest to merit a particular description. In the case, however, of the comet which
appeared in March, 1843, we are permitted to contemplate phenomena of a somewhat different character from those presented by the great comet just
described, but in which there is fully as much of the
very highest interest.

On the morning of the 28th of February, 1843,
soon after sunrise, a brilliant object was seen only
two or three degrees from the sun, by numerous
observers in different parts of the globe.   It was
seen at the Cape of Good Hope a little before sunset, and at several places in the New England States
as early as half past seven o'clock in the morning.
In the early part of the afternoon it was seen by the
naked eye as a bright light, resembling a dagger in
form, and presented at once the appearance of a

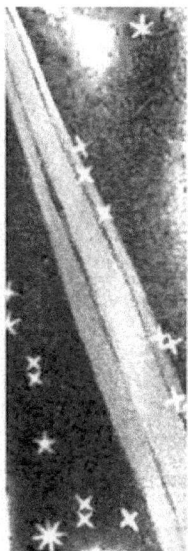

very brilliant comet. The length of the head and
tail was about equal to twice the diameter of the
sun, or about one degree. The head is described at
this time as appearing circular, when viewed with
the naked eye; its light equal to that of the moon
at midnight, in a clear sky; and its apparent size
about one-eighth the area of the full moon. The
train shone with a paler light, gradually diverging
from the nucleus, and fading away by degrees in
the intensely bright sky. It was observed at Wood-
stock, Vermont, with a telescope of small magnifying power, and is described as presenting a distinct
and most beautiful appearance,—exhibiting a very
white and bright nucleus, and a tail divided near
the nucleus into two separate branches. At six
minutes after three o'clock, on the afternoon of the
same day, the distance of the nucleus of the comet
from the sun was measured by Captain Clark, at
Portland, Maine. The distance of the nucleus from
the sun's farthest limb was found to be 4 degrees,
6 minutes, and 15 seconds of arc; or 3 degrees, 50
minutes, and 43 seconds from the centre. The head
of the comet, and also every part of the tail, were
then described as being as well defined as the moon
on a clear day. The nucleus and tail presented the
same appearance, and resembled a perfectly pure
white cloud, without any variation, except a slight
change near the head, just sufficient to distinguish

the nucleus from the tail at that point. In fact, the
nucleus appeared so dense that it was probable that
it might have been visible on the sun's disc, in case
it had passed between this and the observer. The
entire length of the comet, as already remarked,
was at this time about twice the diameter of the
sun; and since the distance of the comet and sun
from the earth were nearly the same, as will soon
be evident, we shall find that the length of the
nucleus and train combined was 1,700,000 miles.

The comet was observed also by means of a sex-
tant, in Mexico, by Mr. Bowring, from nine o'clock
in the morning of the 28th of February, until near
sunset, and its altitude repeatedly measured. It
was seen on the same day at Boulogne, Parma, and
Genoa, in Italy; and it is reported also that it was
seen on the 27th of February, by Captain Ray, at
Conception, in South America. It was then east of
the sun, and distant from the nearest limb about
one-sixth of his diameter, or about 130,000 miles.
It was seen at Pernambuco, in Brazil, and in Van
Dieman's Land, on the 1st of March; and the following day it was very generally observed in the
tropical latitudes in the northern hemisphere. On
the 3d, the head was so far disengaged from the
rays of the sun, as to be visible above the horizon
for a short time after sunset; and when seen through
a telescope of considerable power, it presented the

appearance of a planetary disc, from which were divergent rays in the direction of the tail. The tail is said to have appeared double, consisting of two principal lateral streamers, making a very small angle with each other, and divided by a comparatively dark line, and to have been about 25 degrees in length. It seemed, however, to be prolonged, on the north side, by a divergent streamer, making an angle of at least 5 degrees with the general direction of the axis, and plainly distinguishable as far as 65 degrees from the head. A similar, but fainter prolongation, is said to have been visible on the south side. The representations of the comet at this time, show it to be highly symmetrical, and give it the appearance of a vivid cone of light, with a dark axis, and nearly rectilinear sides, inclosed in a fainter cone, the sides of which curve slightly outwards.

On the 4th of March, the comet was seen at New York; and on the evening of the 5th, it was generally noticed throughout the United States. From this date until the beginning of April, it presented a most magnificent appearance in the western heavens, and is said to have been so brilliant in some places as to throw a strong light on the sea. The nucleus, however, did not remain brilliant for any considerable portion of time. On the 3d of March it was as bright as a star of the first magnitude; on the 11th it was not brighter than a star of the third magnitude; and on the 19th and 20th it had become so faint that it could not be discerned without the aid of a telescope. The tail still preserved its brilliancy, and remained visible as a great beam of nebulous light, until the first week in April.

The first regular observation of the comet, with a suitable instrument, was made at the Cape of Good Hope on the 3d of March, after which it was regularly observed until its final disappearance—the last recorded observation having been made at Berlin, on the 15th of April. It was not seen in England till the 17th of March, and then only as an immense stream of light, without any appearance of a nucleus. The first observation of the nucleus in Europe was made at Rome and Naples on the 17th of March, and it was subsequently observed at all the observatories on the continent. The opposite diagram represents the comet as it appeared to the naked eye in the tropical latitudes, early in March.

The greatest length of the tail was on the 5th of March, when it was about 70 degrees, and was observed to be slightly curved. Its breadth was a little more than a degree. In the tropical latitudes the tail was about sixty degrees in length, and the nucleus shone forth with almost unexampled lustre, creating the most profound sensations of astonishment and admiration wherever it was seen. On the 11th of March, as seen at Calcutta, the tail was observed to have shot forth a lateral ray or train, nearly twice as long as the regular one, but fainter, and making an angle of about 18 degrees with its direction, on the southern side. This projection of a secondary train to such an enormous length — it having been nearly 100 degrees — was not observed either before or after this date. The length of the tail did not vary much after it had attained its maximum, but remained nearly stationary, in this respect, until the first week in April, when it ceased to be visible to the naked eye, owing to the increased distance of the comet from the earth. The apparent length and size of the train, although enormous, hardly serve to give a definite idea of its absolute size. The greatest apparent length was at the time when the comet was nearest the earth, but when it was still distant more than 80,000,000 miles. Let us suppose, therefore, that the length of the tail was found to be 60 degrees, or about 115 times the apparent diameter of the sun, and we shall find its absolute length to be 86,000,000 miles. At the time of the discovery of the comet, in the daytime, the tail was turned nearly toward the earth, and consequently appeared brighter, yet shorter, than would otherwise have been the case. Had it been at right angles to the plane of the orbit of the earth, supposing its apparent length to have been the same as actually observed, its actual length would not have exceeded 2,000,000 miles. It is not certainly known whether the comet had a train of any considerable extent previous to its perihelion passage. It is probable, however, that it had; but from what is now known of the development of the tails of comets, it could not have been near as large as immediately after the perihelion passage. The length of the tail for each day during which the comet was visible, will serve to show the progress of its development, and is exhibited in the following table:

Date.	Length of tail in miles.	Date.	Length of tail in miles.
Feb'y 28,	35,000,000	March 15,	105,000,000
March 1,	55,000,000	" 17,	106,000,000
" 2,	70,000,000	" 19,	107,000,000
" 3,	82,000,000	" 21,	108,000,000
" 4,	91,000,000	" 23,	109,000,000
" 5,	96,000,000	" 25,	108,000,000
" 6,	99,000,000	" 27,	107,000,000
" 7,	100,000,000	" 29,	105,000,000
" 8,	101,000,000	April 1,	102,000,000
" 9,	102,000,000	" 3,	99,000,000
" 10,	102,800,000	" 5,	95,000,000
" 11,	103,000,000	" 7,	89,000,000
" 13,	104,000,000		

The breadth of the tail at the extremity varied from one to three millions of miles. The length of the streamer seen at Calcutta on the 11th of March, must have been about 150,000,000 miles; and if we suppose this to have been projected in a single day, or perhaps in much less than 24 hours, it seems almost, if not actually impossible, to form any conception of the forces which must have operated to produce such a velocity of projection of material substance through space. It is certain, at least, that the velocity must have been such as no other natural phenomenon is capable of exciting; and that in cases similar to this we have to deal with forces incomparably superior in energy to any with which we are acquainted, and with matter whose inertia must be extremely insensible.

The head of the comet, when compared with the dimensions of the tail, was small. It was hardly perceptible even on the 28th of February, when its brilliancy was greatest — the comet presenting the appearance of a mere point, with a divergent mass of matter in the form of a tail. For this reason its cometary nature was at first doubted by those unacquainted with such objects. A few days later the head, although faint in comparison with the tail, could be very distinctly seen, but remained visible to the naked eye only for a very few days. During the last days on which the comet was observed with large telescopes, it was almost impossible to distinguish the nucleus, even while the train was yet brilliant to the unassisted eye. It was for this reason that Bessel, in speaking of the appearance of the comet, remarked that it seemed to have exhausted its head in the manufacture of its tail — a

11

remark which, when we come to consider the theory of the formation of the tails of comets, will have due significance. The diameter of the head of the comet was probably about 8000 miles, and the absolute diameter of the nebulosity surrounding the head was about 30,000 miles. These results, in connection with the dimensions of the tail, already given, will serve to exhibit the enormous magnitude of this celestial visitor.

This comet did not exhibit any indication of a solid nucleus, but, on the contrary, considering the size and brilliancy of the head as compared with the tail, it seems certain that the entire comet was composed of nebulous matter in a state of extreme tenuity — a supposition which the elements of its orbit will furnish the most unequivocal testimony to confirm. However great, then, the volume of the comet may have been, the absolute quantity of matter was inconsiderable, when compared with even the smallest planets of our system. Indeed, it is highly probable that the very smallest members of the group of small planets between Mars and Jupiter, those whose diameters do not exceed a few hundred miles, greatly exceed in weight the entire mass of the comet. The question may therefore arise, as to what is due the intensity of its light when seen at noonday, on the 28th of February, in the full blaze of the sun. For this the elements of the orbit furnish a ready and complete method of solution.

The elements of the orbit of the great comet of 1843 may well be said to be among the most remarkable of those which have hitherto been recorded. The first elements which were computed actually gave the perihelion distance of the comet, measured from the sun's centre, as less than the semi-diameter of the sun, which is a physical impossibility, since the comet was observed after its perihelion passage. A complete discussion of all the observations taken during a period of forty days, made the nearest approach of the centre of the nucleus to the centre of the sun only 390,000 miles, or about 50,000 miles below his surface. Another determination made it 405,000 miles, or about 36,000 miles less than the radius of the sun. Were these results correct, the comet would have plunged directly into the sun in passing the perihelion of its orbit, and would, in all probability, have become absorbed in the solar atmosphere. It would, therefore, have been visible only before the perihelion passage, which took place on the evening of the 27th of February, the day previous to its discovery. The average distance of the comet in its perihelion, obtained by the astronomers who were engaged in computations respecting its orbit, was about 510,000 miles. If we adopt this distance,—and there is the

greatest probability of its correctness,—we conclude
that the comet approached the sun's surface within
about seventy thousand miles, or within one-sixth
of his radius. We perceive, therefore, that the
comet almost literally grazed the sun's disc — a fact
which serves to explain the great brilliancy of the
nucleus at this time.

It is well known that the intensity of both the
light and radiant heat of the sun, at different distances from that luminary, increase proportionally
to the spherical area of the portion of the visible
hemisphere covered by the sun's disc. In the case
of the earth, this disc ·has an average diameter of
a little more than half a degree; but, could an observer have been stationed on the comet, when in
the perihelion, the apparent angular diameter of the
sun would have been nearly 125 degrees. We shall
therefore find, from the principle just stated in reference to the distribution of light and heat by the
sun, that the amount of light and heat received on
equal areas of exposed surface of the earth and
comet, would be in the ratio of one to forty-nine
thousand. Now, it has been determined by experiment, that, by collecting the rays of the sun to such
an extent that the intensity of his heat and light as
received at the earth is increased one thousand
times, the heat produced is sufficient to melt any
mineral substance known on our earth. With this

heat iron is melted in a few seconds, and such minerals as cornelian, agate, and rock crystal. The heat of the sun, as received at the comet, must have been at least forty-nine times more intense than the heat thus generated. The great comet of 1680 was distant from the sun's surface about 150,000 miles, when in its perihelion, or more than double the corresponding distance of this comet; and yet it was computed by Newton that the former was subjected to an intensity of heat 2000 times that of red-hot iron. How much more intense, then, must have been the heat received by the comet of 1843! Such a temperature would actually have converted our earth into vapor; or, if any substance retained a solid form, it would have been in a state of the most intense ignition. On the morning of the 28th of February, the comet must have been literally red hot, although evidently in a gaseous state; and, from its appearance as observed, it retained its heat for several days after its perihelion passage, having presented a peculiar fiery appearance. In the tropical latitudes, and more especially in the vicinity of the equator, it was described as resembling a stream of fire from a furnace. It may be conjectured, therefore, that in case the comet had originally a solid nucleus, it was converted into vapor at this approach to the sun; while the intensity of the heat, and the effect of its operation, were such that the

11 *

comet retained this gaseous form during the entire
period of its visibility in 1843.

The comet did not, however, remain long in a
position exposed to such an enormous range of temperature. Its velocity at this time was not less than
370 miles in a second of time, or 1,332,000 miles
per hour, which would be sufficient, if undiminished,
to cause the comet to make an entire revolution
around the sun in two hours and one-third. This
velocity was so great only for a small portion of
time, and rapidly diminished as the comet proceeded
in its orbit. It was still such that during the twelve
hours immediately preceding the perihelion passage,
and the twelve hours immediately following it, the
comet passed over an arc of its orbit amounting to
290 degrees, or, in other words, performed more
than three-quarters of its circuit around the sun.
During the brief period in which it was visible, it
had described 173 degrees of its orbit, measured
from the perihelion; while to describe the next
7 degrees will require many years, at least, and perhaps many centuries. It thus appears that in about
two hours the heat of the sun, as experienced by
the comet, would have been reduced to about one-
fourth its maximum amount, and that afterwards
the diminution was proportionally greater. In this
manner, then, we may explain the various changes
which were observed to take place in the brilliancy

of the comet, and also the extreme intensity of its
light at the time of discovery.

In computing the elements of the orbit of this
comet, it was found by many astronomers to give
strong indications of ellipticity, while others were
led to conclude that the entire series of observations
could be represented within the limits of their probable errors, by supposing the orbit to be a parabola.
Encke obtained a hyperbolic orbit, whose eccentricity, however, differed but little from that of the
parabola. But it should be remarked that before
any calculations had been made, it was asserted
that, in the year 1668, the tail of an immense comet
had been seen at Lisbon, at Boulogne, at several
points in Brazil, and elsewhere, occupying nearly
the same position in the heavens, and at the same
season of the year. Cassini observed the tail at
Boulogne, on the 10th of March, 1668; and from
this determined that the head was in the immediate
vicinity of the sun. The tail was about 50 degrees
in length, and was so brilliant that its reflected trace
was easily distinguished on the sea. This brilliancy,
however, lasted only for a few days; and on one or
two occasions the head could be barely distinguished
in the twilight, but as a very dim object. The comet
of 1668 was not very accurately observed, and its
orbit was unknown: yet the strange coincidence of
similar situation, season of the year, and physical

appearance, excited a very strong suspicion of identity, thus assigning to the comet a period of exactly 175 years. By estimating the position of the head from the observed direction of the tail, Henderson was enabled to determine an approximate orbit for the comet of 1668; and he found, strangely enough, that the elements closely resembled those of the comet of 1843. Assuming, then, their identity, and using the corrected elements of the latter comet, it was found that the accordance was still better; and it was, therefore, announced that these comets not only exhibited the same appearance, but pursued very nearly the same path, both real and apparent. Again, it has been remarked that comets are recorded to have been seen in the years 268, 442, 791, 968, 1143, 1317, and 1494, which may have been previous returns of the comet now under consideration, since a period of 175 years would,—disregarding the planetary perturbations, which, from the position of the orbit, must necessarily be small,— make it appear in the years 268, 443, 618, 793, 968, 1143, 1318, and 1493.

Such are the reasons for supposing the great comet of 1843 to have a periodic time of 175 years; but it should be noticed that a similar comet was also seen in 1689. It was not observed in Europe, but was seen at Pekin and in the southern hemisphere, where the tail was estimated at upwards of

60 degrees in length. It was observed from the 11th to the 23d of December; and from the observations, rude as they are, its orbit has been computed. Its perihelion distance was small, and the places may be closely represented by elements similar to those of the comet of 1843. Those who have felt inclined to consider these comets identical, have assigned a period of 21 years and 10 months, which gives seven revolutions between 1689 and 1843. By this means the identity of the comets of 1668, 1689, and 1843, is supposed to be clearly established; and the fact that the comet has not been observed at intermediate returns, is accounted for by the position of its orbit, which is such that the comet is best seen in the southern hemisphere, and may pass unobserved in the northern hemisphere. If we adopt this hypothesis, the comet may be expected to reappear about the end of 1864 or beginning of 1865, but will be seen only in the southern hemisphere.

Some astronomers have gone still further, and have supposed the comet to have a period of only 7 years, assigning the unfavorable situation of its orbit as a reason for its not having been seen at its successive returns. This, however, seems wholly inadmissible, since a comparison of all the best observations militates against the hypothesis of an ellipse of short period. They may all be represented by a parabola within the usual limits of the errors of

I

cometary observations, although an ellipse of about
180 years represents them a little better. Still later
determinations of the orbit, in which the probable
errors of the observations are investigated, and the
planetary disturbances duly introduced, increase the
period; and in one instance it is found to be an
ellipse corresponding to a period of revolution of
over 500 years. It is, therefore, extremely doubtful
whether the comets of 1668, 1689, and 1843, are
identical, and the doubt can be removed only by
another appearance of the comet. The greatest
probability is in favor of the identity of the comets
of 1668 and 1843 alone, thus establishing a period
of 175 years within a day or two, more or less; yet
in this case there exists great uncertainty, and such
that we can hardly venture to predict its reappearance in the year 2018.

The most then which we can say respecting the
great comet of 1843, is that the prodigious length
of its tail and its small perihelion distance, both
being such as have never before been observed,
invest it with peculiar interest; and many years may
elapse before another shall appear which, in so short
a space of time, will excite an interest so universal,
or a sensation so profound. We have been minute
in the description of the details of its appearance,
in order that the degree and kind of interest which
is attached to the comets, even by astronomers, in

the present state of the science, may be understood, and also for the purpose of exhibiting the important results which the exact calculations of modern astronomy afford us.

The next comet which became visible to the naked eye, was discovered by Colla, at Parma, on the 2d of June, 1845. It had a very beautiful tail, about three degrees in length. It has been supposed to have a period of about 249 years, making it identical with a comet which appeared in 1596. Another bright comet was discovered by Hind, at London, on the 6th of February, 1847. It passed its perihelion on the 30th of March, at a distance from the sun of about four millions of miles, and was so bright that it was observed at London at mid-day, when only a few degrees from the sun. A comet was also discovered by Miss Mitchell, at Nantucket, Mass., on the 1st of October, 1847, in the vicinity of the north pole of the heavens, which subsequently increased in brilliancy, and on October 6th became visible to the naked eye. It continued to increase in brilliancy for several days, when it was obscured by bright moonlight. The tail was very faint, and did not exceed two degrees in length. It passed the perihelion of its orbit on the 14th of November.

On the 10th of June, 1853, a faint telescopic object, which was subsequently found to be a comet, was discovered by Klinkerfues, at Göttingen. It con-

tinued to increase in brilliancy, and on the 7th of
August began to be faintly visible to the naked eye.
On the 20th of August it was as bright as stars of
the third magnitude, and ten days later it equalled
in brilliancy the brightest stars of the first magnitude. On the 31st of August it was observed at
Olmütz in the day time, when only twelve degrees
from the sun; and on the 2d, 3d, and 4th days of
September it was observed at noon, although only
seven degrees from the sun. It was also seen at
Liverpool on the 3d of September, about noon, by
means of a telescope. The first appearance of a tail
was about the beginning of August, when it had
attained a length of nearly a quarter of a degree.
On the 22d of August it was nearly two degrees in
length, and for a little more than a week continued
to increase, until it attained a maximum length of
fifteen degrees. During this time it presented a
very beautiful appearance in the western heavens,
soon after sunset, to the naked eye, and its motions
were watched with no small degree of interest. It
passed its perihelion, or point of its orbit nearest
the sun, on the 1st of September; but it is not known
to have been observed in the northern hemisphere
after the 4th of September. It was seen at the Cape
of Good Hope on September 11th, and was observed
until the 11th of January, 1854. It was also observed at Santiago, in Chili. from the 16th of Sep-

tember till the 7th of October, and at New Zealand from September 14th to October 10th. About the middle of September the tail was five degrees in length, from which time its brightness rapidly diminished, till it became invisible to the naked eye. The orbit is found to be a parabola, and, consequently, the comet will never reappear to the inhabitants of our earth, but will pass on in its ceaseless wanderings to visit other suns and systems.

On the evening of the 2d of June, 1858, a faint nebulosity was discovered by Donati, an Italian astronomer, at Florence, in the constellation *Leo*, which the observations of a few days proved to be a comet. It was then very far distant from both the earth and sun, and had also a very slow motion as seen from the earth. The comet was seen at nearly all the observatories in Europe in less than a week after the announcement of its discovery. The news of the discovery did not reach America until about the 1st of July, since it was not generally announced in Europe until near the middle of June. The comet, however, had already been independently discovered by Tuttle, at Cambridge, Massachusetts, on the 28th of June; by Parkhurst, at Perth Amboy, New Jersey. on the 29th of June, and by Miss Mitchell, at Nantucket, on the 1st of July, each observer being at the time unaware of its previous discovery in Italy. It was immediately observed at all the

12

principal observatories in the United States, and henceforward observations became general on both continents. As soon as a sufficient number of observations had been taken, the elements of its orbit were computed, and it was found that the comet was distant from the earth, at the time of its discovery by Donati, about 240,000,000 miles, and that it was slowly approaching the sun, and receding from the earth. The distance of the comet from the earth attained its maximum about the middle of June, when it remained stationary for a few days in respect to its geocentric distance. It soon began to approach the earth, and it was found that its brilliancy would be subsequently increased nearly three hundred times. On account of the slow motion of the comet, and its great distance, it was found to be difficult to determine the time of its perihelion passage with any considerable precision. Indeed, the computations of different astronomers exhibited a great difference in this element; and it was not till near the middle of August that its future course could be satisfactorily ascertained. It was then found that the comet would pass the point of its orbit nearest the sun on the 29th of September, that it would become visible to the naked eye early in September, and that during the latter part of this month, and the first half of October, it would present a magnificent appearance in the western heavens, soon after sunset. It was found also that the comet would be so situated with respect to the earth and sun, that it would not only be visible in the evening after sunset, but during the greater part of September it would rise several hours before the sun, and, consequently, would be favorably situated for observation in the morning.

The predictions were fully realized. The comet was distinctly visible to the naked eye on September 10th, and continued to increase rapidly in brilliancy until about the 5th of October, when it had attained its maximum in this respect. It ceased to be visible in northern latitudes, except near the equator, about the 20th of October, but was observed in the southern hemisphere from the beginning of October to the middle of May, 1859. The first appearance of a tail was noticed about the 20th of August, with the aid of a telescope, and early in September it was very distinctly noticed to be curved. Between the 10th and 25th of September, the comet increased rapidly in brilliancy, and the train was lengthened proportionally. The comet was now rapidly approaching both the earth and sun, and under circumstances, as we shall subsequently notice, peculiarly favorable to the development of the tail. Another fortunate circumstance to be noticed in this connection, was that the moonlight, about the time of the greatest brilliancy of the comet, ceased

to interfere with the observation of the minutest
details of its form, and contributed largely to en-
hance its general appearance. The return of the
moonlight just before it ceased to be visible above
the horizon in the south-west, and when its brilliancy
was speedily on the decline, caused its decrease and
final disappearance to be fully as sudden as had
been its increase and maximum intensity of light.

At the time of its perihelion passage the comet
was distant from the sun about 55,000,000 miles.
Its nearest approach to the earth took place on the
11th of October, at which date it was distant from
us about 50,000,000 miles. The following table
gives the distance of the comet from the earth and
sun, and also its hourly motion in its orbit, for different dates during the period of its visibility in the
northern hemisphere.

Date.	Distance from sun in miles.	Distance from earth in miles.	Hourly velocity in miles.
June 2, . . .	214,000,000	240,000,000	65,000
July 1, . . .	175,000,000	241,000,000	72,000
Aug. 1, . . .	128,000,000	221,000,000	84,000
Sept. 1, . . .	82,000,000	160,000,000	105,000
" 15, . . .	64,000,000	116,000,000	119,000
Oct. 1, . . .	56,000,000	66,000,000	128,000
" 11, . . .	61,000,000	52,000,000	123,000
" 21, . . .	71,000,000	67,000,000	114,000

The brilliancy of a comet, as seen from the earth,
varies inversely as the product of the squares of its
distance from the earth and sun respectively; and

it results, therefore, from the foregoing table, that this comet was 200 times brighter on the 1st of October than at the time of its discovery. The brightest stars in the heavens are about 100 times brighter than those which can be barely distinguished by the naked eye. This fact may serve to give an idea of the comparative brilliancy of the comet at these two dates.

The tail of this comet is certainly the most remarkable feature to be considered. At the time when the comet was nearest the earth, it could be distinctly traced to a distance from the nucleus of nearly 70 degrees, and was curved in the form of a sabre. On the 20th of September, it was first observed to be bifurcated, or divided into two branches or streams. The dark space between these two streams of nebulous light was directly behind the nucleus, and gave it the appearance of the shadow of the latter obscuring the light of the tail. The southern portion of the tail was much more brilliant than the northern portion; and the difference was such that, in the twilight, the latter could not be distinguished, so that the tail appeared single. The length of the tail, at this time, was about 5 degrees. On the 24th of September the length of the tail was more than 7 degrees, and the curvature began to be very plainly exhibited. On the 27th its length was 13 degrees, and presented the phenomenon of a

12 *

streamer, or supplementary tail, issuing from the
convex side of the principal one, and nearly in the
direction of a straight line drawn from the sun
through the nucleus. From this date until the 10th
of October, the tail increased rapidly in brilliancy
and length; and from the phenomena which it presented, there is every reason to believe that it was
directly connected with the changes which, as we
shall notice, were continually taking place in the
matter composing the envelope and nucleus. At
times, rays or jets of light were seen streaming in
different directions from the centre; and coruscations were seen, precisely as in the case of several
comets which have already been described. On the
5th of October there were two streamers distinctly
visible, one of which was more than fifty degrees in
length, corresponding to upwards of 50,000,000
miles. At this time, the comet was so bright that
not only the nucleus, but also a portion of the envelope and tail could be seen, by means of a telescope, before sunset. It was visible to the naked
eye in the bright twilight; and, when viewed with
a telescope, a secondary envelope could be distinctly
noticed. At the point where the curvature of the
tail began to be most strongly marked, several short
streamers could be seen; and occasionally there
seemed to be transverse bands in the tail, nearly a
half a degree in breadth, with clear, well-defined

outlines, and closely resembling the auroral streamers, with the exception only that they retained their
relative positions in the tail.   On the 10th of October, the date on which the tail had attained its
greatest dimensions, these slightly diverging bands,
— the point of divergence being situated between
the sun and the nucleus, — alternating with dark
spaces, could be readily distinguished, and in reality
gave the comet a most beautiful appearance.   They
were, on an average, five degrees in length, and
from a third to half a degree in breadth.   The opposite diagram represents the comet as it appeared at
this time.

The apparent length of the tail was necessarily
affected by the moonlight which prevailed at two
different times during the period of visibility, and
we are therefore unable to determine the precise
variations in its length from day to day.   The following gives the length of the tail for successive
dates, as observed at the Observatory at Cambridge,
Massachusetts:

Date.	Length of tail in miles.	Date.	Length of tail in miles.
Aug. 29, 1858	14,000,000	Sept. 30, 1858	26,000,000
Sept. 8, "	16,000,000	Oct. 2, "	27,000,000
" 12, "	19,000,000	" 5, "	33,000,000
" 17, "	10,000,000	" 6, "	45,000,000
" 23, "	12,000,000	" 8, "	43,000,000
" 24, "	12,000,000	" 10, "	51,000,000
" 25, "	17,000,000	" 12, "	39,000,000
" 27, "	18,000,000	" 15, "	14,000,000
" 28, "	26,000,000		

The discrepancy which seems to exist in the measurements from the 17th to the 25th of September, inclusive, is to be attributed to the influence of the moonlight; and any other slight variation from a regular increase which may be apparent, is probably due to a hazy or unfavorable state of the atmosphere at the time when the measurements were taken.

The breadth of the tail at the extremity was found, on the 25th of September, to be 1,500,000 miles. On the 30th it was 3,000,000 miles, and on the 10th of October 10,000,000 miles. The streamers were from 10,000,000 to 50,000,000 miles in length; and in one instance, at least, their breadth at the extremity was upwards of 1,000,000 miles.

The nucleus of the comet was observed to undergo the most rapid changes that can be readily conceived of, not only in form, but also in size. On the 19th of July, its diameter was found to be 5600 miles; August 30th, it was 4660 miles; and September 24th it was only 1030 miles. On the 5th of October, it was certainly less than 540 miles in diameter; while on October 18th it had increased to 900 miles. On the 8th of September, the diameter of the nucleus was ascertained to be 2000 miles; and in immediate contact with it was an intensely brilliant nebulosity, having a diameter of about 3000 miles, while the diffused light was estimated to extend

40,000 or 50,000 miles in the direction of the sun.
About two weeks later, it was noticed that there
was interposed between the nucleus and the sun, an
obscure, crescent-shaped outline, within which the
light was unequally distributed, and which had a
strangely confused and chaotic look. The nucleus
at this time was singularly brilliant, and appeared
elongated on the upper side—a phenomenon which
was supposed to indicate the existence of some internal disturbing force. Outside of the nucleus
there was a bright envelope, whose vertex was in
the direction of the sun, and about 6000 miles distant from the nucleus. This envelope was surrounded on its outer margin by a dark band, sepa-
rating it from a second and less brilliant envelope,
whose vertex was distant from that of the first one
about 7500 miles. This again was encircled by a
dark arch, outside of which there was an atmosphere
of faint and diffused nebulosity, which gradually
faded away, and became invisible at a distance of
about 40,000 miles from the nucleus. These bands
could be traced through an arc of more than 200
degrees, around the nucleus; but extended further
into the train on the brighter or upper side with
reference to the horizon.

The phenomena which these successive bands or
envelopes presented, was of the most instructive
and interesting character. They were observed in

succession to disengage themselves from the nucleus, then gradually to expand and recede, and finally to disappear in the formation of the tail. So plainly was this exhibited, that each envelope could be traced through all its successive stages; and thus have the means been furnished for investigating in detail the mysterious processes by which the train is thrown out from the nucleus, under the action of forces whose nature and mode of operation will be fully explained when we come to treat of the theory of the physical constitution of comets, and of the formation of their tails. In this way, also, we are enabled to account for the rapid decrease of the diameter of the nucleus as the comet approached the sun, when the matter of which it is composed was driven off to form the tail; and, also, the increase in the size of the nucleus as the comet again receded. On the 3d of October, the nucleus appeared to be divided into two distinct parts,—one, however, being much brighter than the other. These were separated by a distinctly dark opening, and were distant about 4000 miles. No change was observed at this time in the relative positions and form of the envelopes, although the tail seemed to have been sensibly affected — the bifurcation being much more distinct, and broader in comparison with the breadth of the tail.

The opposite diagram is a telescopic view of the

nucleus and part of the tail of the comet on the 3d
of October, 1858.

The successive envelopes are here exhibited precisely as they appeared when the comet was seen
through a powerful telescope. The division of the
nucleus is also indicated, and the dark space sepa-
rating the tail into two parts. The development of
these envelopes, as already remarked, was extremely
interesting. In most cases they were observed to
escape from the vertex of the nucleus; but this was
by no means general. They were also seen escaping
from the sides of the nucleus, and have been described as an escape of jets of luminous gas, which
streamed off like light spray thrown up against an
opposing wind and driven before it — a description
which is certainly very aptly conceived. The entire
distance between the vertex of the nucleus and the
inner surface of the outer envelope, varied from ten
to twenty thousand miles. It was noticed, also, that
just before an eruption took place, the brilliancy of
the nucleus was perceptibly greater than either immediately before or after that event; and the variations were so rapid as to indicate almost the exact
instant at which the eruption took place. In order
to show the progressive motion of the envelopes
from their point of origin, the following distances
from the nucleus to the vertex of one of these envelopes, for each day during the period of its trans-

formation, as observed by Bond, at Cambridge, Mass., are given:

Date.	Distance of Vertex from nucleus in miles.	Date.	Distance of Vertex from nucleus in miles.
1858, Sept. 27, . . .	3,500	1858, Oct. 6, . . .	10,100
" " 29, . . .	6,000	" " 8, . . .	12,400
" Oct. 2, . . .	7,500	" " 9, . . .	13,200
" " 4, . . .	8,900	" " 10, . . .	14,100
" " 5, . . .	9,550		

It appears, therefore, that the daily motion of the vertex of the envelope, relative to the vertex of the nucleus, varied from 300 to 1200 miles. This rate of motion was found to coincide very closely with that of the other envelopes at the same date. The motion of the envelopes, it might further be remarked, varied with the position of the comet with respect to the sun. The manner in which these envelopes were driven off, and the evident tendency of each towards a condensation around a central axis, together with the fact that the axes manifested a disposition to diverge from the sun, has led to the conjecture that these collections of nebulosity were in reality a group of new comets in process of formation. It is therefore unfortunate, that the circumstances were such as to preclude the possibility of following them to a more complete development.

We may thus comprehend the peculiar interest which attaches to the great comet of 1858. The

long period during which it was visible before it had
attained its most magnificent proportions, had enabled astronomers to be prepared to observe all the
phenomena of the changes of its physical appearance with the greatest facility. Other comets have
appeared which furnished phenomena of precisely
the same character, and perhaps on as grand a scale,
but the state of astronomical science was such that
their importance was not fully understood, and consequently the observations were not so refined and
extensive. Under such circumstances, we may declare, without hesitation, that the comet now under
consideration has been of more essential service to
science, and otherwise of more universal interest,
than any other which has ever been seen. The
splendor of its appearance in October elicited the
most profound admiration throughout the civilized
world; and although, like its predecessors, it was
assigned its share in the cause of any unusual events
which may have taken place simultaneously with
its appearance, yet it is certain that the general
diffusion of knowledge which characterizes the
present age, operated very effectually to counteract
all superstitious fears, which, otherwise, it would not
have failed to excite.

The computations which have been made respecting its orbit, show that the period of its revolution
is about 2000 years. The following are some of

13   K

the results which have been obtained by different
astronomers :

Bruhns,	. . . . .	2102 years.
Watson,	. . . . .	2415 "
Löwy,	. . . . .	2495 "
Brünnow,	. . . . .	2470 "
Newcomb,	. . . . .	1854 "

The uncertainty of the period arises from the
great eccentricity of the orbit; and since the observations made in the southern hemisphere during
several months after its disappearance in the northern, have not been employed in the determination
of these results, this uncertainty may be materially
reduced in subsequent computations. The disturbance produced in its motion by the attraction of the
planets during the period of its visibility, have also
been neglected, the introduction of which may affect
the results slightly.   It is probable, however, that a
more complete investigation will give a period differing but little from 2400 years.   The comet must
therefore be distant from the sun, when in the aphelion or most remote part of its orbit, no less than
34,000,000,000 of miles, or about one six-hundredth
part of the distance of the nearest fixed star. Great
as is this distance, the comet will still be obedient
to the controlling influence of the great central
body of our system; and after having performed its
long journey of more than two thousand years, it

will again return to greet the inhabitants of our
earth.

We have thus briefly described some of the most
remarkable comets which have appeared during the
period of authentic history. To notice all which
appeared would not only have been tedious and
uninteresting, but also would have required more
space than could properly have been devoted to this
branch of our subject. The number of comets which
have appeared since the birth of Christ, in each
successive century, is as follows: First century, 22;
second, 23; third, 44; fourth, 27; fifth, 16; sixth,
25; seventh, 22; eighth, 16; ninth, 42; tenth, 26;
eleventh, 36; twelfth, 26; thirteenth, 26; four-
teenth, 29; fifteenth, 27; sixteenth, 31; seventeenth, 25; eighteenth, 64; nineteenth (till 1860),
114. Many of these have been reappearances of
the same comet; and although the total number
here given is 651, yet we are not to understand that
there has been 651 different comets observed during
this time. Of these, we have described those which
were remarkable for their brilliancy, and whose
periods of revolution have not been exactly determined. The periodic comets, properly so called,
will be described when we shall have explained the
manner in which the orbits of these bodies are
determined.

The number of comets here given comprises only

those whose appearances have actually been recorded, while it is perfectly certain that many have
visited the sun unperceived. By comparing the
positions of the orbits of those comets whose elements have been determined, it is found that, unlike
the planets, they are distributed over the entire
surface of the heavens, without preference of any
one region to any other. Again, the points where
they pass through the plane of the earth's orbit, are
found to be uniformly distributed in every direction
round the sun. The points where they pass nearest
the sun are also distributed uniformly round that
body, and their perihelion distances are such as
leads to the supposition of their uniform distribution
through space. If now we assume this to be actually the case, since these bodies can only be seen
from the earth when within the orbit of Mars, we
may be able to form an estimate of the probable
number of comets in space. From considerations
such as these, and by a process of reasoning which
it is not necessary to explain, Arago concludes that
at least *seven millions* of comets are enclosed within
the limits of our system. But a small portion of
these, comparatively, can ever be seen from the
earth; and although this hypothesis may seem unreasonable, yet there is every reason to regard it as
a very approximate estimate of the number of these
wonderful bodies.

CHAPTER III.

ORBITS OF COMETS — COMPUTATION OF AN ORBIT — EPHEMERIS OF A
COMET — PERIODIC COMETS — HALLEY'S COMET — ENCKE'S COMET —
BIELA'S COMET — FAYE'S COMET — DE VICO'S COMET — BRONSEN'S
COMET — D'ARREST'S COMET — TUTTLE'S COMET — WINNECKE'S COMET
— LEXELL'S COMET OF 1770 — OTHER PERIODIC COMETS — RELATION
SUPPOSED TO EXIST BETWEEN THE COMETS OF SHORT PERIOD AND
THE ASTEROID PLANETS — POSSIBILITY OF A COLLISION BETWEEN
A COMET AND PLANET — PROBABLE APPULSE OF TWO COMETS.

As soon as it was ascertained that the comets have
no connection with the earth, but move in the region
of the planets, the investigation of their orbits, and
their relation to the other known bodies of the universe, became a problem of the very highest interest.
The illustrious Kepler endeavored to represent their
observed places, by supposing them to move in
straight lines with a variable velocity—a supposition
which was subsequently adopted by several astronomers; and computations were made, in accordance
with it, respecting the motions of some of those
comets which had previously appeared. This hypothesis of Kepler, as we shall see, is now known to
be unfounded, yet the real paths pursued by the
comets are such that, for a considerable distance,
they very nearly coincide with a straight line; and

13 *

we may therefore understand how Cassini was enabled to predict the places of a comet, for a few
weeks in advance, with considerable precision.
Hevelius, a few years later, found, by observing the
comet of 1665, that its path was curvilinear, and
that the orbit might be a parabola. He did not,
however, conjecture what would be the position of
the sun with respect to the curve; and it was not
till the appearance of the great comet of 1680, that
this question was definitely settled. It was then
determined, by Dörfel and Newton, that the orbit of
the comet was a parabola, with the sun situated in
a fixed point within the curve, called the *focus*. It
was further determined by Newton, that the comets
form a part of the solar system; that they describe
orbits around the sun in obedience to his attractive
force; and that their motions are in accordance with
the same laws which govern the motions of the
planets. He also explained a method of determining
the orbit from positions of the comet observed at
the earth. This was the first attempt to reduce the
motions of the comets to numerical computation,
in a manner which could be fully investigated by
mathematical analysis.

In order to effect a determination of the size and
position of the orbit of a comet, Newton proposed
two different methods. In the first he supposes
that a small portion of the orbit may be regarded

as a straight line, described with a uniform motion, and that its segments, intercepted by straight lines drawn from the earth to the comet, at the times of the respective observations, are proportional to the intervals of time between these observations. In this manner the ratio of these segments becomes known; and since the portion of the orbit included between the extreme observations is a straight line, its projection on the plane of the earth's orbit is also a straight line, divided similarly to the orbit itself by the projections of the straight lines joining the places of the earth and the comet. It would seem, therefore, to be a very simple process to find the projected orbit; but it happens, unfortunately, that this method is one which, in the case of the orbit of a comet, admits of almost an infinite number of solutions, and, consequently, is so nearly indeterminate that no definite results can be obtained. The other method proposed by Newton was more general and complete; and, as far as theory is concerned, is wholly unexceptionable. It is found, however, that when actually applied in the determination of an orbit, it leads to computations excessively complicated and laborious, and consequently has never been generally adopted.

The theory of the determination of the orbits of comets remained for many years where Newton had left it; and astronomers were still attempting, but in vain, to discover a method by which an orbit might be determined with accuracy, without making it necessary to have recourse to such extended numerical calculations. Failing in every attempt at a direct solution, they were compelled to adopt a method of trial and error, and thus, by successive hypotheses, to approximate finally to an orbit which would nearly represent the observations. Boscovich, however, undertook to solve this difficult problem directly; and the solution which he gave is remarkable as being the first in which the velocity of the comet in its orbit was regarded as one of the essential conditions. His method, as might be expected, is excessively complicated; and from the multiplicity of both algebraical formulæ and graphical operations which it presents, it has never been employed in the actual computation of an orbit.

The next attempt to solve this problem was made by Lambert; and among the formulæ which he obtained relative to the motion of a body in a parabola, there is one extremely remarkable, on account of its elegance, which gives the time of describing an arc of the orbit in terms of its chord and the two lines drawn from its extremity to the centre of the sun. This formula is still employed in every method which is in use for the solution of this problem, since it affords a very easy means of correcting the hypo-

thetical values which it is necessary to substitute in
the equations of the problem. He also showed how
it may be determined at once whether the distance
of the comet from the sun is greater or less than the
distance of the earth from the sun, thus enabling
the computer to abridge, very materially, the preliminary computations. The method of Lambert
was subsequently modified by the researches of Lagrange and Laplace; and the results finally arrived
at by these illustrious geometers are now adopted,
for the most part, as the basis of all analytical solutions. Still another method was proposed by Olbers
for finding the orbit of a comet moving in a parabola,
in which, as in the case of those last mentioned,
only three different places, as seen from the earth,
are required. He supposes that the chord which
joins the places of the comet at the epoch of the
first and third observations, is divided by the line
drawn from the comet to the sun, at the time of the
second observation, into segments proportional to
the intervals of time between the observations.
This method is the simplest and most direct which
has been devised; but gives accurate results only
when the intervals of time betwen the dates of the
observations are very nearly equal. In those cases
where the inequality of the intervals is considerable,
and especially when the comet is further from the
sun than the earth, it gives very unsatisfactory re-

sults; and for this reason Legendre has devised a
method suitable to all such cases, in which—instead
of assuming that the chord of the arc of the orbit
between the first and third observations is divided
by a line drawn from the comet to the sun at the
time of the second observation, in the ratio of the
elapsed times—he finds an expression for the value
of each of the segments in terms of the time. This
last method may be employed with success in cases
where that of Olbers fails, but the numerical calculations are much more complicated and tedious.

We have thus stated, somewhat in detail, some of
the various methods which have been devised for the
solution of the problem which requires the orbit of
a comet to be determined by means of its successive
positions observed from the earth; and what is here
given will be sufficient to show that it is one of the
most difficult problems of astronomy. If we already
knew, and could subject to calculation, the causes
which originally determined the motions of the
heavenly bodies, we could assign, at once, the data
for a complete solution of this problem without the
aid of observation. But as these are, and must ever
remain, unknown, we have no means of arriving at
the desired results, except by the method just explained. Again, if the observer could be situated
at the centre of the sun, the real and apparent motions of the comet would be identical, and the deter-

mination of its orbit would be a process extremely
simple.   But since the observer is stationed on the
earth, which is itself in motion in an eccentric orbit,
it may readily be perceived that it is necessary to
refer all positions, as seen from the centre of the
earth, to the corresponding places referred to the
centre of the sun, in order to arrive at the desired
result, and this it is which makes the problem
difficult.

We have already remarked that it was discovered
by Newton that the comets were obedient to the
same laws which regulate the motions of the planets
belonging to the solar system; and we are, therefore,
enabled to adopt in this case, also, those beautiful
laws of planetary motion discovered by Kepler.
These laws are three in number, and are as follows:
1st. The heavenly bodies revolve around the sun in
conic sections whose common focus is the sun; 2d.
The radius-vector, or line joining the planet or
comet with the sun, describes equal areas in equal
times, and, consequently, in unequal times areas
proportional to the times; 3d. The squares of the
periodic times are to each other as the cubes of the
mean distances of the bodies from the sun.   These
three laws, the direct result of the law of universal
gravitation, together with three observed positions
of the comet, are all the data required in finding the
magnitude and position of its orbit, and its position

in the orbit at any given time. But we have here spoken of a class of curves called conic sections, and it may not therefore be improper to explain their character in this connection.

If we take a cone—which may be conceived of as being a solid formed by the revolution of a right-angled triangle around the perpendicular side as a fixed axis — and pass planes through it at different degrees of inclination with respect to the axis, the intersections of these planes with the surface of the cone will give us one or all of the three curves known as *conic sections;* namely, the *ellipse, parabola,* and *hyperbola.* The plane which passes through the cone so as not to cut the base, but only the slant surface, gives an ellipse, a curve which, next to the circle, is the most simple. Those which cut the base will mark out on the slant surface of the cone segments of either the parabola or hyperbola, depending simply on the position of the plane with reference to the axis. This explains why these curves are termed conic sections; but their true character will be much more readily comprehended by a more familiar illustration.

If we take a fine thread and fasten its extremities at two points on a plane surface, the distance between these points being less than the length of the thread — which will thus lie loosely between the points — and if now this thread is kept tightly stretched by means of a pencil which is made to move round the entire circumference permitted by the looseness of the thread, the curve thus described will be an *ellipse*, or, as it is sometimes called, an *oval*. Now it is easy to be seen that since the thread has retained the same length, and has been kept tightly drawn out during the description of the curve, the sum of the lines drawn from any point of the curve to the two points at which the thread was fastened, will be the same for every such point in the curve. The two points within the curve at which the ends of the thread were fixed, are called the *foci* of the ellipse, and in the case of the elliptic orbit of a comet the sun is placed in one of these points.

Let us now suppose that the distance between these points is increased, the length of thread remaining the same; and, as may be readily seen by experiment, we shall obtain a curve similar to the one already traced, only that its form has been altered, while the length of the line joining any point in the circumference with the new foci, remains as in the previous case. By continuing to increase the distance between the foci, using always the same thread, we shall find that the successive ellipses will be more and more flattened or elongated, until finally they will differ but little from a straight line. If, on the contrary, the distance between the foci is successively

14

shortened, the ellipses formed will gradually become less and less flattened, and will scarcely differ from a circle. If the foci be made to coincide, then will the curve traced be a circle, and for this reason a circle is sometimes said to be a species of ellipse. The line which is drawn through the foci and terminating in the curve at either extremity, is called the *major axis* of the ellipse; while that which bisects this at right angles, terminating also in the curve at opposite sides, is called the *minor axis*. The ratio of the major axis to the line joining the foci is called the *eccentricity*. The eccentricity will therefore be greatest when the foci are at their greatest distance, and least when they coincide, as in the case of the circle.

Having thus explained the method of projecting an ellipse, and the general properties of the curve, we will proceed to illustrate the construction of a parabola. For this purpose let a ruler be placed in any desired position on a plane surface, and remain fixed in that position. Another ruler is now placed at right angles to this, but capable of being moved along it. The latter must be so arranged, that in moving along the former they shall always form a right angle; or, in other words, one shall continue perpendicular to the other. Let us now take a string equal in length to the perpendicular edge of the movable ruler, and fix one end of this string at any

point in the plane surface on the same side of the
fixed ruler on which the movable one is placed, the
other end being attached to the latter at the outer
extremity of the perpendicular edge. If, then, by
means of a pencil, we commence at the point where
the string is fastened to the movable ruler, and move
it slowly towards the fixed ruler, always keeping the
string close to the edge of the former, and at the
same time move the latter along it in the direction
of the fixed end of the string — passing this point,
and continuing until the pencil arrives again at the
other extremity of the string, the curve thus traced
will be a parabola. By varying the distance of the
fixed point of the string from the fixed ruler, a great
variety of parabolas may be traced out. The fixed
ruler is called the *directrix* of the parabola, and the
point within the curve at which the string is made
stationary, is called the *focus*. In the case of the
orbits of the comets, the sun is placed in this, the
only focus of the parabola. It may thus be perceived that the parabola is constructed in such a
manner, that the distance between the focus and any
point of the curve is equal to the perpendicular distance between the same point and the directrix.

Next, to describe an hyperbola. On a plane surface draw two straight lines at right angles and bi-
secting each other, the horizontal line, however,
being the longest. Then take two points on the

longest of these lines, one on each side of the per-
dicular line and at equal distances from it. Place a
ruler so that one end of it may revolve around one
of these points — being kept closely pressed to the
plane surface—and to the other end attach a string,
which is also to be fastened at the other extremity
to the fixed point taken on the opposite side of the
perpendicular line from that around which the ruler
revolves. The ruler is now revolved slowly around
its axis, while the string is stretched by means of a
pencil, commencing at the point where it is fastened
to the ruler, and moving it towards the other end of
the same, until it finally passes the horizontal line,
when it will move back again along the ruler to the
point of beginning. The string must be kept close
to the edge of the ruler at the point where the pencil is placed, and in the same manner for every point
in the curve which will thus be traced out on the plane
surface, and which is called an hyperbola. The
fixed points which have been taken in the horizontal
line are called the *foci* of the hyperbola, and in the
case of the orbit of a comet moving in an hyperbola,
the sun is placed in the focus which is within the
curve. The point at which the fixed lines first drawn
intersect, is called the *centre* of the hyperbola. The
point where the curve crosses the horizontal line is
called the *vertex* of the hyperbola, and the distance

between the vertex and the point of intersection of
the fixed line, is called the *semi-transverse axis* of the
hyperbola. That part of the perpendicular line
which is limited by an arc of a circle described
around the vertex of the hyperbola as a centre, with
a radius equal to the distance of the focus from the
centre, is called the *conjugate axis* of the hyperbola.
The ratio of the semi-transverse axis to the line
joining the centre with the focus; or, which is the
same, the quotient which arises from dividing the
latter by the former, is called the *eccentricity;* and it
will be perceived, from what has just been stated,
that this ratio will always be greater than unity.
In the ellipse, it follows, from the same principles,
that this ratio or quotient resulting from a division
of the line joining the foci, by the major axis, will
always be less than unity. The eccentricity of the
ellipse may, therefore, vary between the limits 0 and
1, the eccentricity of the parabola is 1, while that
of the hyperbola may have any value greater than
this.

In this way these curves are found to bear an intimate relation to each other, and in order to exhibit at once their characteristic differences, together
with their general resemblance, the following diagram is given:

14 *   L

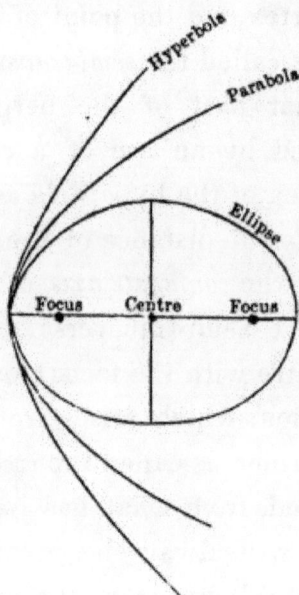

The distance between the focus and vertex is the same for each of the above curves; and this distance, in the case of the orbit of a comet, is called the *perihelion distance*.

The curves which we have just described we have supposed to have been traced out with a short piece of thread; but curves may be imagined as being constructed in accordance with the same laws, and having therefore the same properties, traversing vast and inaccessible regions of space, and in which those dimensions which we have represented perhaps by tenths of an inch, are replaced by millions of miles. Such are the orbits of the comets; and in one of these curves every body which traverses our system is constrained to move, in obedience to the law of universal gravitation. The planets, however, although permitted by that law to move in similar paths, are found to revolve around the sun in orbits which are purely elliptical, and in which the eccentricity is generally very small. They are found also to move in a series of orbits at distances increasing in a regular progression, confined to limits of only a few degrees on either side of the orbit of the earth, and in the same direction in which the earth pursues its annual course around the sun. It might thus seem that an accordance so wonderful, and an order so admirable, could not be fortuitous; and not being enjoined by the law of gravitation, must be ascribed to some additional character appertaining to this law, with which we are unacquainted. But the comets show this beautiful law exemplified and obeyed in all its comprehensive features; and thus sweep around the sun in all the varied forms of ellipses, parabolas, and hyperbolas, in all planes, at all distances, and indifferently in both directions in reference to the motion of the planets.

The comets are usually seen from the earth when near the sun, and consequently when not far distant from the perihelia of their orbits. It will be evident, therefore, by an inspection of the diagram on page 162, that, since the three conic sections nearly coincide for a considerable distance on either side

of the vertex,—which corresponds to the perihelion
of the orbit,—it will be difficult to determine, from
the small segment of the orbit observed, whether
the comet moves in an ellipse, a parabola, or an
hyperbola. If the centre of gravity of the comet
could be observed with the utmost precision, this
would not be difficult; but, as a general fact, the
observer only estimates the point of greatest condensation of light, which he assumes to be the centre of gravity, and the place of which in the heavens
he determines by reference to the fixed stars in its
immediate vicinity. We are thus left in doubt in
regard to the exact form of the orbit, unless it happens that the comet is favorably situated for its
determination.   It is therefore customary among
astronomers, when a comet has made its appearance
unpredicted, to compute its orbit at first on the supposition that it is a parabola; and then, by computing its place in advance, find from a comparison
of the actual observations, whether this hypothesis
is the correct one. Should it be found to be impossible to represent the observed positions of the
comet by a parabola, an ellipse is next computed;
and when this also fails, recourse is had to the
hyperbola, which, provided the previous computations are correct in every particular, will not fail to
represent the observations within the limits of their
probable errors. The results of these computations

give the *elements* of the orbit of the comet. These
elements are six in number — of which two refer to
the nature and magnitude of the orbit, three to its
position in space, and one to the position of the
comet in its orbit at a given time. They are, therefore, the following : 1st. The perihelion distance, or
distance between the focus and vertex of the curve,
the sun being placed in the focus ; 2d. The eccentricity, which in the case of the parabola becomes a
known quantity ; 3d. The inclination of the plane
of the orbit of the comet to the plane of the earth's
orbit, or to what is called the *ecliptic ;* 4th. The position of the line in which the orbit intersects the
ecliptic ; 5th. The position of the greater axis of the
orbit in space ; and 6th. The time at which the
comet was in its perihelion, or its angular distance
from this point at any stated epoch. In order to fix
the position of the line of intersection of the orbit
and ecliptic, — which is called the *line of nodes,* —
and also the position of the greater axis of the orbit,
it is necessary to adopt some standard point of
departure, to which all angular measurements may
be referred. The point used by astronomers is that
in which the sun, in its apparent annual course,
passes from the southern to the northern hemisphere
of the heavens, and is termed the *vernal equinox.*
When the comet is in the line of nodes, in the act
of passing from the south to the north side of the

ecliptic, it is said to be in its *ascending node*, and the
position of the line of nodes is determined by the
angular distance of the ascending node from the
vernal equinox. This angular distance is called the
longitude of the node. The position of the greater
axis of the orbit is determined by the angular distance between the vernal equinox and the perihelion
point, projected orthographically on the ecliptic.
This distance is called the *longitude of the perihelion.*

Such are the *elements* required for a complete
knowledge of all the past and future circumstances
of the motion of a heavenly body revolving around
the sun, in obedience to the law of universal gravitation. We have already remarked, that if we knew
the exact conditions under which the comets were
first put in motion, we could assign the proper values
of these elements at once, without the aid of observations; but since these conditions are unknown,
we are compelled to reverse the process, and by
means of the effects produced to seek out the causes
which originally operated. To accomplish this,
three observed positions of the comet are required.
These observations should be taken at intervals of
at least one day, and should be referred to the centre of the earth. These three observations give us
the direction of the three visual rays drawn from
the earth to the comet at the times of the observations, and in the prolongation of which the latter

must necessarily be found. The corresponding places of the sun are known either by observation, or by calculation from the solar tables. It remains then to find a parabola, ellipse, or hyperbola, having its focus at the centre of the sun, and cutting the visual rays in points, the intersections of which correspond to the number of days between the observations—the motion in the orbit being in accordance with Kepler's laws. In order to effect this, it is necessary, in the first place, to refer the positions of the comet as seen from the earth, to those in which it would appear if seen from the centre of the sun; or, in other words, to transform the *geocentric* places of the comet into *heliocentric* positions. This is accomplished by computing a series of triangles, formed by joining the successive places of the earth and comet with the sun, and by joining the places of the earth with those of the comet. The solution of these triangles, together with Lambert's theorem, in the case of a parabola, furnish the means of passing from the earth to the sun. As soon as the heliocentric places are found corresponding to the dates of the observations respectively, it becomes an easy process to determine the position of the orbit in space, since, at the centre of the sun, the real and apparent motions of the comet are the same. The transformation which gives the heliocentric longitude and latitude of the comet, gives at the same time its distance from the centre of the sun, or what is called its *radius-vector*, corresponding to the three observations. We shall thus have three points of the curve given; and by finding the ellipse, parabola, or hyperbola, as the case may be, which shall pass through these three points, a segment of the orbit will be given. Now, the properties of these curves have been so completely investigated, that if a segment, — no matter how small, — is given, the entire curve may be readily traced out. This enables us to determine the perihelion distance and the eccentricity of the orbit, in the case of the ellipse and hyperbola, the latter being already known in the case of the parabola. The time of passing the perihelion is then found, by means of Kepler's laws, from the times of the observations.

Such is a concise explanation of the method of determining the *elements* of the orbit of a comet from three observations; and, although the process, as here stated, may seem to be by no means difficult, yet it should be understood that this problem, if properly solved, is one of the most intricate and laborious of all the problems which the science of astronomy affords. Moreover, a detailed account of the operations to be performed in making the computations, would require a higher degree of mathematical knowledge, in order to be fully comprehended, than is possessed by those for whom these explanations are intended; and what has already been stated will be sufficient to exhibit, in an intelligible manner, the general character of the problem, and also of its solution.

When the elements of the orbit of a comet have been determined, it becomes an easy process to predict its position in the heavens for any future date. It is customary among astronomers, as soon as the orbit is known, should the comet continue visible, to compute a series of places, for each successive day, during the period of its visibility. These places are indicated by means of the computed right ascension and declination of the comet— the former being the angular distance between it and the vernal equinox, measured on the ecliptic towards the east; and the latter, its angular distance north or south from the celestial equator. The distance of the comet from the earth and sun, respectively, is also determined at the same time. The positions of the comet thus arranged constitute an *ephemeris;* and, in case the elements have been accurately determined, all the various changes in the brilliancy of the nucleus or head may be predicted for many months in advance.

In computing an ephemeris it is necessary to find the position of the comet in its orbit for a given epoch, by means of the time of perihelion passage, the eccentricity of the orbit, and the perihelion dis- •

15

tance. This being known, the heliocentric place
of the comet is readily found; and then, by an easy
transformation, the geocentric right ascension and
declination may be determined. In practice, however, a course slightly different from this is pursued.
The position of the comet in its orbit is determined
as here explained; but instead of finding the heliocentric place, and from it the geocentric place, the
position of the comet with respect to three imaginary planes, at right angles to each other, and
having their point of common intersection in the
centre of the sun, is investigated. These planes
are called co-ordinate planes, and the point of common intersection is called the *origin*. The three
lines of intersection of the planes are called *co-ordinate axes*, and the distances of the comet from the
origin, measured in the direction of these lines, are
called its *rectangular co-ordinates*. The rectangular
co-ordinates of the earth are then found by means
of the solar tables, which, being subtracted from
those of the comet, give the rectangular co-ordinates
of the comet referred to the centre of the earth.
In making these transformations, the plane of the
equator is usually adopted as the fundamental plane,
and is supposed to coincide with one of the co-ordinate planes. The ecliptic, also, may be adopted as
the fundamental plane; but, in this case, the final .
.places of the comet would be its latitude and lon-

gitude in the heavens, instead of its right ascension
and declination. When the geocentric rectangular
co-ordinates have thus been computed, they may be
transformed at once into polar co-ordinates, thus
giving the apparent position of the comet on the
celestial vault. It should be remarked, however,
that there are certain small corrections which must
be added to these results, in order to find the exact
place of the comet, but which are not of sufficient
importance to be noticed in this connection. The
ephemeris is now compared with the actual observations of the comet; and should any discrepancies be
exhibited, the elements are corrected so as to reduce
these differences within the limits of the probable
errors of the observations. In this way, therefore,
new and more accurate elements are obtained, and
an exact ephemeris is then computed precisely as
before.

When a new comet has been discovered, it is observed for a few days, and as soon as three places
are obtained, separated by an interval of time of
sufficient extent, the elements are computed under
the supposition that the orbit is a parabola. An
ephemeris is then computed from the elements, and
compared with the entire series of observations, including an interval of time of the greatest possible
extent. Should it be found, from this comparison,
that the observations are not satisfactorily repre-

sented, the elements are corrected as above explained; and in case discordances are still found to exist, it becomes necessary to abandon the hypothesis of a parabolic orbit, as incompatible with the motion of the comet. An ellipse is now computed, and compared with the observations; and if, after having been finally corrected, it is still impossible to satisfy the entire series of observations, recourse is had to the hyperbola. The comparison thus instituted between the observations and the computed places resulting from the hypothesis of a parabola, an ellipse, or an hyperbola, respectively, will indicate the precise character of the orbit. The final effect of these successive hypotheses on the relation of the comet to our system, may be readily conceived. A comet which moves in a parabola cannot make, like one which moves in an ellipse, a succession of revolutions around the sun. It enters the system in some definite direction, but as proceeding from almost an indefinite distance. As soon as it arrives within the sensible influence of solar gravitation, the effects of this attraction are manifested in the curvilinear form which its path begins to assume, and which gradually increases as its distance from the sun decreases. When it reaches its perihelion, the attractive force, and consequently the curvature of its path, have attained their maxima, and the comet has now also acquired its greatest velocity of translation. This extreme velocity, by virtue of the inertia of the moving mass, produces a centrifugal force which counteracts gravitation; and the body, having passed its perihelion, begins to retreat, and pursuing a similar path to that in which it previously moved, only in a reversed order, it finally passes out of the system in nearly a straight line, and parallel to that in which it entered. It is evident from this, that a comet which moves in such an orbit, can visit the solar system but once, unless it may happen that in passing through some other adjoining system, its course is so changed by the attractive influences which may there be in operation, that it again retraces its way, in its majestic flight, towards our system, but in a direction entirely different from that in which it first entered.

In the case of an hyperbolic orbit, the circumstances of the motion of the comet with respect to the sun would be nearly the same as results from parabolic motion. The only difference would be, that in the parabola the comet enters the system in nearly a straight line, and issues from it in a straight line parallel to this; while in the hyperbola these straight lines are divergent, and not parallel. The comet, in either of these cases, would not have a periodic character. It might be said to be analogous, in this respect, to one of those occasional meteors

15 *

which are seen to shoot across the firmament never again to reappear. The body, arriving from some distant region, and coming, as would appear, fortuitously within the solar attraction, is drawn from its course into an hyperbolic or parabolic path, which it is seen to pursue ; and escapes from the solar attraction, issuing from the system, and, for aught we know, never to return. The phenomenon, in each case, might be said to be occasional, and, in a certain sense, accidental, and the body could not be said to belong properly to the system. So far as relates to the comet itself, it may be said that the phenomena consists in a change of the direction of its course through the universe, influenced by the temporary action of solar gravitation upon it.

Such are the circumstances under which a great majority of the comets which have visited our system, have been found to move ; and they may on this account be regarded as a connecting link between the various systems of the universe. But in the case presented by a comet moving in an elliptic orbit, the relation between the comet and our system is much more intimate, and the interest and physical importance of the body transcendently greater. In this case the comet possesses a periodic character, making successive revolutions like the planets, and returning to its perihelion at regular intervals of time. Unlike those comets which move in parabolas or hyperbolas, it would not be an occasional visitor to our system, connected with it by no permanent relation, and subject to solar gravitation only accidentally and temporarily. On the contrary, it is to be regarded as being as permanent, if not as strictly regular, a member of the system, as any of the planets, although invested with an extremely different physical character. This being the most plausible view of the cometary motions in general, and one which would seem to have impressed, at once, every one who has studied the phenomena of these wonderful bodies, it may perhaps seem strange that no attempt was made until recently to determine the orbits of the comets under the direct supposition of elliptic motion. But it should be remembered that those which are known to be periodic move in ellipses of great eccentricity; and since they are observed only when near their perihelion, it becomes almost impossible to decide in reference to the nature of the orbit. As already remarked, the observations are not absolutely exact; and making due allowance for the probable errors of observation, even when the comet has been visible for a long period of time, it is generally found that its visible path may be represented with almost equal fidelity by either of the three curves in which these bodies move. In all such cases, it is impossible to infer, from the observations alone, whether the comet

belongs to the class of hyperbolic or parabolic
bodies, which have no periodic character, or to the
elliptic, which has.

When the period of revolution does not exceed
100 years, the periodic time may be found within
a year of the true time from the observations made
during two or three months. In this case the ellipticity is so marked that the observations will not
fail to indicate it. When the period is only about
five or fifteen years, the observations of three months,
provided they are numerous, will give the orbit so
approximately that the next appearance of the comet
may be predicted within a few days of the observed
time. The method of determining the elements of
the orbits of the heavenly bodies has been carried
to so high a degree of perfection, that by combining
many observations made at different places nearly
simultaneously, or within a few days of each other,
thus forming what are called normal places of the
comet, almost the exact form and magnitude of the
orbit can be found directly. But although, in the
case of a periodic comet, the ellipticity is plainly
exhibited, and its value very closely approximated,
yet the unavoidable errors of observation are still
sufficient to render the period of revolution to a
limited extent uncertain. Sometimes this uncertainty may amount to several weeks, and again it
may not exceed a few days or perhaps a few hours

in the time of the return of the comet to its perihelion. It depends necessarily on the character and
number of the observations employed, and also upon
the interval of time during which the comet was
accurately observed, or in a position favorable for
good observations.

It often happens that the character of periodicity
itself, which of course belongs only to elliptic orbits,
supplies the means of surmounting the difficulties
just enumerated as affecting the exact determination
of the periodic time of a comet, from the observations made during a brief interval. If any observed
comet has an elliptic orbit, it must return again to
its perihelion after having completed one revolution,
and it must also have been visible on former returns
to that position. It is evident, therefore, that not
only ought the comet to be expected to reappear in
future at regular intervals, but that its previous returns to its perihelion would be found, in case it had
been observed, by searching among the recorded
appearances of such objects for any, the dates of
whose appearances might correspond with the supposed period, and whose apparent motions might
indicate a real motion in an orbit, identical, or nearly
so, with that of the comet which may be under consideration. Now it is known that, in most cases,
the elements of the orbit undergo comparatively
slight changes during one, or even two or three suc-

cessive revolutions. If the force of gravity of the sun operated alone, and the comet was not subjected to the attractions of the various planets of the system, the elements would continue invariable. The comet would reappear after each successive revolution at exactly the same point as seen from the sun : would follow, while visible, exactly the same arc of its orbit; would move in the same plane, inclined at the same angle to the ecliptic, the nodes retaining the same places; and would arrive at its perihelion at exactly the same point and after exactly equal intervals of time. The disturbing action of the planets is usually small, especially when the inclination of the plane of the orbit of the comet to the plane of the ecliptic is considerable. In general, this disturbance, which is called the planetary perturbations, will not change any of the elements more than one or two degrees; while in many instances the variations in the position of the orbit in space with respect to the equinox and ecliptic—these being the elements which are most affected — will not exceed half a degree. It may happen, however, — the orbit of the comet being very eccentric, and on that account nearly intersecting the paths of several planets—that the disturbing action of these planets may, in the aggregate, entirely, or nearly overcome the principal effect of the predominant force of the sun; and the orbit of the comet at two different returns will be so completely changed that it would be impossible, from a mere inspection of the elements corresponding to each appearance, to identify it. While, therefore, we may be prepared for the possibility, and even the probability, that the same periodic comet on the occasion of its successive reappearances, may, in passing to and from its perihelion, follow a path differing to some extent from that which it had followed at previous returns; yet, as we have already remarked, such differences cannot, except in rare and exceptional cases, be very considerable. For the same reason the intervals between its successive periods, though they may differ, cannot be subject to any great variation.

The elements of the orbits of all the comets which have appeared during the last century, and also of many comets which had previously appeared, have been computed and recorded. As soon, therefore, as an unexpected comet has made its appearance, and the elements of its orbit have been determined, an examination of these recorded elements is made, in order to detect, if possible, any which have a sufficient resemblance to those of the comet in question, to warrant the supposition of their identity. Should it be found, on making such a comparison, that there are no elements on record which satisfy the required conditions of identity, we may conclude that the comet is a new one, and must have recourse

to actual observations for the determination of the
precise character of its motion. The approximate
elements already known will enable us to conjecture
at least the probable influence of the principal
planets, and we are thus enabled to make due allowance, in the comparison, for any discrepancies in the
elements which might result from this cause. In
the case of a parabola, there are five elements to be
compared ; and since, considering a limited number
of comets, the chances are so small that a near resemblance of all the elements may be accidental, we
may conclude without hesitation, in all cases where
such a resemblance is detected, that the comets are
identical, and that whatever differences are found
to exist, must be due to the attraction of the planets
and to the unavoidable errors of the observations.

The identity of a comet thus determined does not
always furnish the period of its revolution. It is
not known whether the comet has made but one
revolution during the interval which has elapsed, or
whether it may not have returned several times unperceived. This question must be decided by actual
computations from the observations. If it is found
that, by assuming an eccentricity less than that
which would result from the supposition of a single
revolution, the observations may be more completely
represented, it will be evident that the comet has
made more than one revolution between the epochs

of the two sets of elements. By successive approximations, conducted in the same manner, or perhaps by direct computation of elliptic elements from the observations, this question may be definitely settled, and the periodic time may thus be accurately determined.

Nor is the fact that no identity can be found by a comparison of the elements of a comet which may be under consideration, with those which have been recorded, in itself sufficient evidence of its not having a fixed period of revolution. It may have returned always unperceived, or it may have been seen but not observed sufficiently for the determination of its orbit. Prior to the time of Kepler and Galileo, these bodies were often visible for a long period of time, but were not observed, except as objects of curiosity, or as portending the greatest imaginable disasters. Indeed, we find that historians have mentioned, and even described, their appearances, and in some cases have indicated the principal constellations through which such bodies passed, although no observations of their apparent places have been transmitted, by which any close approximation to their actual paths could be made. If, therefore, the elements of a comet are known so approximately as to give its period of revolution within limits of a few years, and it is found that comets have been thus described at epochs which

16

might correspond with its previous appearances, the identity is possible. In such cases it is only necessary to reduce the modern elements to the ancient epoch, and by computing the path of the comet, find whether it would have presented the phenomena, both in respect to brilliancy and apparent motion, which were then observed. Should they be found to agree, it is certain that the comets are identical.

The resemblance in magnitude, form, and splendor, which may be found to exist between comets which have appeared at different epochs, does not afford any reasonable basis for supposing them to have been successive appearances of the same comet. In modern times more reliable data are furnished for the solution of this problem; while in more ancient times, the only circumstance generally recorded was the appearance of the object, accompanied, in many instances, with details bearing evident marks of exaggeration. It is only, therefore, in cases where the computations of the elements, from modern observations, indicate unequivocally the elliptic form of the orbit, and its approximate periodic time; or, where a distinct and close resemblance between two different systems of elements for comets at two different epochs, respectively, is found to exist; that the precise period of revolution of a comet can be made known, except in the case already referred to, in which the period does not exceed fifteen years. It may thus be perceived how astronomers are enabled to make such wonderful predictions in regard to the future movements of some of these objects, as have been made from time to time. In no other science can be found so marvellous a series of phenomena foretold. The interval between the prediction and its fulfillment has sometimes exceeded, and will often exceed, the limits of human life; and one generation bequeathes its predictions to another, which is to be filled with astonishment and admiration at witnessing their literal accomplishment.

We have thus explained in a concise, yet sufficiently extended manner, the process by which astronomers are enabled to predict the past and future movements of a comet, from observations made only during a brief period in which it is visible. In the explanations here given, it was, of course, impossible to enter into a detailed account of all the various methods which may be employed in the solution of this problem. We have given that which is most simple and readily understood, and which is sufficient to demonstrate to those unacquainted with the abstruse and intricate mathematical investigations which are effected by the professional astronomer in making these calculations, the theory of the solution, thus showing conclusively that the final results arrived at, though

sometimes startling, are of the most positive
character.

In the preceding chapter we gave a description
of some of the most remarkable comets which have
appeared — reserving, however, until now, a description of the phenomena connected with the
periodic comets, or at least those which have either
been identified at more than one appearance, or are
known to have periods of revolution not exceeding
one or two hundred years. We shall therefore proceed at once to give an account of this latter class
of comets, in the order in which their periodic character has been discovered.

It might, indeed, be expected that comets moving
in elliptic orbits of small dimensions, and consequently having short periods, would have been the
first in which the character of periodicity would be
discovered. The comparative frequency of their
returns to their perihelia, and the consequent possibility of verifying the fact of their periodic character, together with the distinctly elliptic form of their
orbits, which would be supposed to be made evident
by computation, would afford a firm basis for such
an expectation. But it is found in this case, as has
happened in so many others in the progress of physical science, that the actual results of observation
and research have been directly contrary to such an
anticipation; the most remarkable case of a comet

of large orbit, long period, and consequently rare returns, being the first, and those of small orbits, short periods, and frequent returns, the last, generally speaking, whose periodicity has been discovered.

Soon after the discovery of the law of gravitation by Newton, it was found to confirm completely the laws of Kepler, and also to be constant and universal in every situation in which it could be traced, whether on the earth or in the solar system. This fact gave to the physical sciences generally, and to that of astronomy in particular, a generalization and harmony which had not previously been known. The general law of the motions of the heavenly bodies through space, in connection with the observations of the great comet of 1680, led Newton to conclude that the orbits of the comets must, like those of the planets, be ellipses having the sun in one focus, but far more eccentric; and having their *aphelia*, or greatest distances from the sun, far remote in the regions of space. He adds, however, that he leaves it to others to determine the greater axes of their orbits and their periods of revolution, by comparing comets which return after long intervals of time, with the same orbits. The idea thus advanced by Newton, extraordinary as it may appear, was soon to be strictly realized. Twenty years later, Halley collected all the observations of comets

16 *

which he could procure; and, by means of the method invented by Newton, he determined with great difficulty the elements of the orbits of twenty-four, which, out of 425 of these bodies recorded prior to the beginning of the eighteenth century, were all that had been observed with sufficient accuracy to admit of the determination of their orbits by the method which he employed.

On comparing the elements thus obtained, Halley found that the orbits of the comets of 1531, 1607, and 1682, were nearly the same. He also found that one which had appeared in 1661, followed nearly the same path with one which was visible in 1532. He was, therefore, led to conclude that the former comets were reappearances of the same comet, which would thus have a period of revolution of about 75 or 76 years, and that the latter had a period of 129 years. It may perhaps be proper to add that the latter conjecture has failed to be realized, while the former has been found to be literally true. His belief of the identity of the comets of 1607 and 1682, was still further confirmed by records of more ancient comets, among which he found three nearly corresponding in their periods with the former. These were the comets of 1305, 1378, and 1456. He now declared his opinion that the same comet had appeared at these six epochs; and that, since its period of revolution was a little more than seventy-five years, it might be expected to return in the year 1758. The following table shows the resemblance of the periods which led to this important announcement:

Perihelion Passage.		Interval.		
November	9, 1378,	. . . 77 years,	7	months.
June	9, 1456,	. . . 75 "	2½	"
August	26, 1531,	. . . 76 "	2	"
October	27, 1607,	. . . 74 "	10½	"
September 15, 1682,				

Here was exhibited an inequality in the periodic time amounting to more than two years—a circumstance which, in connection with the fact that there was found to be a variation in the inclination of the orbit of the comet to the plane of the ecliptic, might seem to cast a doubt on the conclusion at which Halley had arrived. It was known, however, that the comet might have been sensibly disturbed in its orbit by the attractions of Jupiter and Saturn at its successive returns; and that this would perhaps account for the discrepancies in its periodic time, which would otherwise appear to exist. The attraction of Jupiter on Saturn was known to affect the velocity of the latter planet; sometimes retarding and sometimes accelerating it, according to their relative position, so as to affect its period to the extent of thirteen days. The influence of Saturn on Jupiter was also known to be of precisely a similar character, although its effect was not more than half as great. Under such circumstances, Halley did not hesitate to declare that, if such heavy bodies as the planets could be disturbed in their motions by their mutual attractions, a comet, — which he considered a mass of vapor, — would be still more disturbed in passing near such large and solid bodies, especially if far removed from the influence of the sun. He concluded, also, that since the motion of the comet in its orbit is so rapid, a very small increase of velocity from any disturbing cause would change the form of its ellipse.

For these and similar reasons, he did not hesitate to attribute the inequality observed in the interval between its successive returns, and also the variation in the inclination of its orbit, to the attractive influence of planets. Moreover, he noticed that in the interval between 1607 and 1682, the comet passed so near Jupiter, that his attraction must have augmented its velocity, and consequently shortened its period. In the same manner he found that, after 1682, its period would be again increased by the influence of Jupiter and Saturn; and he therefore predicted, finally, that the comet would reappear toward the latter part of the year 1758, or about the beginning of 1759. It was certainly a bold analogy, in those days, to attribute a difference of more than a year, in the motion of a comet, to the same cause that occasioned a change of only a few

days in the motions of the planets; and this confi-
dent prediction by Halley may therefore be regarded
as one of the most remarkable events in the history
of astronomy. But, great as was this achievement,
the imperfect state of mathematical science rendered
it impossible for Halley to exhibit to the world an
absolute demonstration of the event which he foretold. It was, therefore, only possible for him to
announce these felicitous conceptions of a sagacious
mind as mere intuitive perceptions, which must be
received as uncertain by the world, — however
strongly he himself may have been impressed with
them, — until they could be verified by the process
of a rigorous analysis.

Such was the prediction of Halley — a prediction
which, considering the state of astronomical science
at the time it was made, must certainly be regarded
as one of the most extraordinary and wonderful
which has ever been made. The distinguished philosopher, sensible of the fact that he could not live
to witness the verification of his prophetic announcement, patriotically expressed the hope that, should
the comet return in accordance with his prediction,
posterity would do him the justice to acknowledge
that it was announced by an English astronomer.

Before the comet had accomplished another revolution, at the completion of which Halley's prediction was to be fully realized, mathematical science

had advanced rapidly to a very high degree of perfection, and men of sublime genius had arisen, who
were able to follow out Newton's theory of gravitation to its legitimate results and consequences. Of
the problems thus presented for investigation there
was one of peculiar importance, and yet of the most
difficult character. One of the first and simplest
results of the theory of gravitation was, that if a
single planet revolved around the sun in free space
— its mass being supposed to be insignificant as compared with that of the sun — it must revolve in an
ellipse, the focus of which must be occupied by the
centre of the sun. But if we now suppose a second
planet to be admitted into the system, then the
ellipticity of their orbits can be preserved only on
the supposition that the two planets have no attraction for each other, and that no physical force is in
operation except the solar attraction. This will
necessarily require that their masses shall be so small,
compared with that of the sun, as to be entirely
neglected. The law of universal gravitation, however, is based on the principle that every particle of
matter, every body in nature, must be mutually
attracted. This may seem to indicate that the elliptic
character of the orbit would be effaced; but it should
be remembered, that in the cases of the planets of
the solar system, their masses are so small compared
with that of the sun, that for any limited time their

orbits may be regarded as purely elliptic. But in course of time their orbits must be slightly changed, depending on their relative masses, those whose masses are least being most sensibly affected. These slight deviations are called the *perturbations* of the planets and comets, or the disturbances due to their reciprocal attraction. The problem thus presented is known as the *problem of the three bodies*, and its extension embraces the effects of the mutual gravitation of all the planets of the system upon each other. We must, however, defer, until a subsequent connection, a complete explanation of the method of solving this problem, and of its more extended applications; stating here only what may be required in order to understand the manner in which the exact time of the return of a comet may be predicted.

The problem of the three bodies, involving the sun, earth, and moon, had already engaged the attention of Euler, D'Alembert, and Clairaut; and from the comparative smallness of the third body considered, namely, the moon, the solution was effected with considerable facility. But in order to apply the problem to the case of the sun, a planet, and a comet, it was necessary in the first place to investigate the analytical formulæ which would represent the respective motions of the planet and comet, before the actual place of the latter, at any given time, could be determined, supposing the elements of the orbit at a fixed epoch to be known. This was attempted by Clairaut, one of the most profound geometers then living; and he endeavored to determine the path of the comet of 1682, when attracted by the sun and disturbed by a planet. In this complicated problem, the disturbing action of one planet only can be estimated at a time, and on this account it becomes necessary to repeat the numerical computations for each disturbing body. Moreover, the disturbing action of a planet can only be computed for an extremely small portion of the orbit at a time, and the sum of these is the whole effect.

It was enough for Clairaut to investigate the complicated formulæ by which these computations were to be accomplished. He was indeed eminently qualified to conduct such an investigation, and the manner in which it was accomplished may be witnessed by the accuracy of the final results. When the analytical part of the problem had been solved, the laborious numerical computations were undertaken by Lalande, an accomplished practical astronomer. The fidelity with which he performed his part will soon be made evident. In this prodigious labor he was assisted by Madame Lepaute, an accomplished lady of Paris, whose name — although, for some unaccountable reason, wholly omitted in Clairaut's Memoir giving an account of the computations — is thus deservedly registered in the annals of astronomical science.

When we consider that the period of the comet is about seventy-five years, and that for two successive periods it was necessary that every portion of its orbit should be calculated separately, as above explained, a general idea of the enormous labor performed by Lalande and Madame Lepaute may be formed. They computed from morning till night, and even late in the evening, without intermission, for a period of six months; having computed the distance of each of the two planets, Jupiter and Saturn, from the comet, and their attraction upon that body, separately for each degree of the orbit, during a period of 150 years. The result was, that in consequence of the attractive influence of these two planets, the period of revolution of the comet would be lengthened nearly two years. It was found that it would be delayed 518 days by the action of Jupiter, and 100 days by that of Saturn, so that it would arrive at its perihelion on the 18th of April, 1759. It was afterwards found, on a revision of the calculations, which had been, toward the latter part, performed too rapidly—in some instances neglecting quantities which were by no means inconsiderable — from fear of being anticipated by the arrival of the comet, that the perihelion passage would take place on the 4th of April, 1759.

17   x

The Memoir which contained an elaborate account
of these investigations was presented to the French
Academy on the 14th of November, 1758, and a
supplementary one, containing an account of the
revision of the latter part of the numerical computations, was presented a few weeks later. In this
Memoir Clairaut explained the unfavorable circumstances under which the computations were made.
He had no observations on which to base the elements of the orbit but those of Apian, which were
far from being accurate, having been made at a time
when little attention was paid to comets. It is probable, however, that if Apian had been sensible of
their future importance, he would have observed
with much greater accuracy. The mass of Saturn
was unknown, and Clairaut was compelled to adopt
a value which has since been found to be erroneous.
He also neglected the disturbing action of the earth,
which was not altogether inconsiderable, since the
comet had passed near it in 1682, and could not take
into account the action of Uranus and Neptune,
which were then undiscovered. The effects of the
action of Mercury, Venus, and Mars, being insignificant, were not computed.

It is not strange, therefore, that Clairaut admitted
that his results might be liable to a small error,
enough, perhaps, to leave an uncertainty of one
month in the date of the return of the comet to its

perihelion. He stated, also, that there might be very many circumstances which, independent of any error either in the methods or process of calculation, might cause the event to deviate more or less from its predicted occurrence. He was well aware that a body which passes to such distant regions of space, and which is invisible for so long a time, may be subjected to forces wholly unknown; such, for example, as the action of other comets, or even of some planet so distant from the sun that it has not yet been seen — a fact which has since been established by the discovery of the planets Uranus and Neptune. Again, Newton and many of his followers believed in the existence of an ethereal fluid throughout all space, although there was no proof of it at that time, and conjectured that it would accelerate the motions of the heavenly bodies without altering the position of their orbits. No one, however, had attempted to estimate its effects till Clairaut endeavored to find what influence it would have on the motions of Halley's comet, — which was the name assigned to the comet of 1682, in honor of the discoverer of its periodic time — and he found that the acceleration would not amount to more than seven and a half minutes.

Such were the circumstances attending the prediction of the exact time of the comet's return by Clairaut; and so positively were his results an-

nounced, and so firmly was he convinced of their
general accuracy, that he felt certain that the comet
would first be visible about the end of 1758, or early
in January, 1759.   There was, however, one circumstance connected with the previous returns of the
comet, which seemed to Lalande to indicate the
possibility of the return of the comet in 1759, without being seen.   In 1456 it occupied a space nearly
seventy degrees in length, and spread terror throughout Europe.   In 1607, its appearance, as described
by Kepler, was that of a star of the first magnitude,
and so trifling was its tail that it was at first doubted
whether it had any.   In 1682 it excited very little
attention, except among astronomers; and supposing
this decrease of magnitude and brilliancy to be progressive, Lalande entertained serious apprehensions
that on its expected return it might escape the observation even of astronomers; and that thus, this
splendid example of the power of science and un-
answerable proof of the principle of universal gravitation, would be lost to the world.   It is indeed true,
that, in 1607 and in 1682, the comet was not favorably situated in respect to the earth and sun to appear with great brilliancy, while in 1456 its position
was exactly the reverse of this.   But still it did not
seem hardly possible that so great a change in physical appearance could be caused by simply the
difference of position with respect to the earth and

sun at the time of its visibility; and it may, therefore, be said, that it affords the very greatest interest
to observe the misgivings of this distinguished astronomer, with respect to the appearance of the body
according to the prediction, in connection with his
unshaken faith in the results obtained. He asserted
that it would, without doubt, return; and that, even
if astronomers failed to see it, they would not, on
that account, be less convinced of its presence.
They would know that the faintness of its light, its
great distance, and, perhaps, even an unfavorable
state of the weather, might prevent its being seen.
The world, however, would find it difficult to believe in his results, and would place this discovery,
which had done so much honor to modern philosophy, among the number of chance predictions.

Under such circumstances, the return of the comet
was awaited by astronomers with the greatest im-
patience and anxiety. They were anticipated, however, in the discovery, by George Palitzch, a peasant,
residing in the neighborhood of Dresden. He first
saw the comet on the evening of the 25th of December, 1758, through a small telescope, and on the
next day communicated the discovery to Hoffman,
who observed it on the 27th and 28th of December.
A few days later it was independently discovered by
an astronomer at Leipsic, who, jealous of his discovery, did not announce it, but gave himself up to

17 *

the solitary pleasure of following the body in its
course from day to day, while his cotemporaries
throughout Europe were vainly directing their
anxious search for it to other quarters of the heavens.
Delisle, a French astronomer, and Messier, his assistant, had been constantly engaged in searching
for it, but an error in the computation of the
ephemeris, by the former, had diverted their attention to a different part of the heavens from that in
which the comet was to be seen; and the consequence
was, that it was not seen by Messier until the 21st
of January, 1759, nearly a month after its discovery
by Palitzch, but without knowing that it had been
already observed.

The news of the discovery now became generally
known, and the comet was observed at various
places in Europe. On its first appearance the nucleus was round and brilliant, surrounded by a vapor
or nebulous atmosphere, but did not exhibit any
indications of a tail. It was rapidly approaching
the sun, and about the middle of February was lost
in the approaching twilight. It passed its perihelion
soon after midnight at Paris, on the 12th of March,
1759, only twenty-three days before the time predicted by Clairaut and Lalande, or within the limits
of probable error which they had assigned to their
results. Laplace has since shown, that if the mass
of Saturn had been accurately known at the time

when the computations were made, the error of the final result would not have exceeded nine days. This, together with the fact that the planets Uranus and Neptune have since been discovered, and that the existence of an ethereal fluid pervading the regions of space has been established by its action on the motions of the comets, enables us to form a true conception of the importance of Clairaut's labors, which had revealed the place of the long-expected wanderer, while yet invisible to the naked eye, on its return after an absence of three-quarters of a century, to crown with triumph the illustrious astronomer who had first foretold its period. What greater exemplification of the transcendant powers of the human understanding could be desired, or what greater exhibition of the achievement of genius? An object which had once caused the greatest alarm throughout all Europe, and which had been supposed, in former times, to have caused some of the greatest disasters which have ever befallen the human race, was now divested of its terrific aspect and nature; and hereafter, as it returned successively in its long journey through space, to be hailed with increasing delight, as affording new and conclusive proofs of the harmony of those beautiful laws which govern the celestial motions.

The comet emerged from the sun's rays toward

the end of March, 1759, and was visible in the
morning just before sunrise; and, had it not been
in the bright twilight, was in the most favorable
position, with reference to the earth and sun, for
being seen in all its splendor. On the 1st of April,
Messier was able to distinguish a tail with the aid
of a telescope; but in no other instance was it certainly observed in Europe. On the 17th of April it
ceased to be visible in the morning; on the evening
of the 29th of the same month it appeared about
the size of one of the largest stars, and from the 3d
of June ceased to be visible. The apparent diminution of the comet in magnitude and brilliancy,
already noticed by Lalande, may here seem to be
positively confirmed as a physical fact, since its
splendor at this return was even less than in 1682.
But it should be remembered that it was seen in
1759 under the most disadvantageous circumstances.
In April, when the tail ought to have been longest,
the comet was far from the earth, and rapidly receding from the sun; and, moreover, was almost
always obscured by the effect of the twilight.
Again, it may be readily understood why a comet
may, at successive returns to our system, sometimes
appear to have a tail, and sometimes to be without
one, according to its position with regard to the
earth and sun. It may happen that, at one return,
the earth may be near the comet at or near the time

of its perihelion passage, when its light is necessarily greatest and its train most extended, and thus
a most favorable opportunity will be afforded for
witnessing its physical appearance; while, at its
next return, the earth being at a remote part of its
orbit while the comet is passing the sun, it may be
seen only with great difficulty, or even become
quite invisible. It is evident, therefore, that no
expectation of a uniform physical appearance of a
comet at successive returns, can be indulged; and
that there is scarcely the slightest chance of ever
being able to recognize even a single one among
the many thousands which are sweeping through
the regions of space, by means of any apparent
similarity of form and brilliancy.

In the southern hemisphere, Halley's comet was
much more favorably situated for observation, in
1759, than in the higher latitudes of the northern
hemisphere. It was observed at the Isle of Bourbon
and at Pondicherry, and the tail was distinctly visible to the naked eye at both places, its length
varying from ten to forty-seven degrees, which
seems to accord fully with the former appearances
of the same body.

Such were the circumstances connected with the
first predicted return of Halley's comet; and its
appearance, in accordance with the prediction, excited the curiosity and wonder of both the learned

and unlearned throughout all Europe. Another
interval of seventy-six years has elapsed since its
appearance in 1759, the same computations have
been performed, only that this time the planet Uranus was known, and the mass of Saturn had been accurately determined; and on the 16th of November,
1835, the comet, true to its predicted course, passed
the perihelion of its orbit. This revolution, however, was performed under circumstances far more
favorable than had ever before occurred. During
the interval between 1759 and 1835, the science of
analysis, as applied to physical astronomy, had
made rapid advances. The methods of investigation
had acquired greater simplicity, and had also been
made more general and comprehensive. Learned
societies had likewise been established in several
cities in Europe and elsewhere; and had stimulated
the spirit of inquiry by a succession of prizes
offered for the solution of problems arising out of
the difficulties which were progressively developed
by the advancement of astronomical science. Among
these questions, the determination of the elliptic
orbits of comets, and the perturbations which they
experience in their course, by the action of the
planets near which they happen to pass, were considered of primary importance. The French Academy, accordingly, offered, in the year 1778, a high
mathematical prize for an essay on this subject,

which called forth a Memoir from Lagrange, in which he formed at once a complete solution and a model for all future investigations of the same kind. This investigation by Lagrange was, however, of a general character; and it remained to apply it to the particular case of Halley's comet, the only one then known whose periodicity had been definitely determined. In 1820, the Academy of Sciences at Turin offered a prize for this application of Lagrange's theory, which was awarded to Damoiseau, a French astronomer; and, in 1826, the French Institute proposed a similar prize, having offered it twice before without calling forth any claimant. This prize was awarded to Pontécoulant, who determined the perturbations of Halley's comet by taking into account the simultaneous action of Jupiter, Saturn, Uranus, and the earth — the comet having, in 1759, passed sufficiently near the latter to experience from it sensible disturbances. He found, finally, that it would pass its perihelion on the 7th of November, 1835; and afterwards, by a revision of the calculations, he predicted that its arrival at that point of its orbit nearest the sun, would take place on the morning of the 14th of the same month.

The same problem was undertaken by Lubbock, Rosenberger, and Lehmann. Damoiseau and Pontécoulant had assumed the orbit in which the comet was moving in 1759 as the basis of their investigations, and had calculated simply the alterations which would be produced upon it by the action of the planets from 1759 to 1835. In his second solution, Pontécoulant corrected the elements by means of the observations in 1759, and consequently obtained more satisfactory results. Lubbock, Rosenberger, and Lehmann, undertook in the first place to ascertain the orbit which it followed in 1759, by means of such observations as had been recorded at that time; and then, by making due allowance for the planetary perturbations, to predict the exact date of its return to its perihelion. It results, therefore, as might be expected, that the time of passing the perihelion in 1835, as found by these astronomers, exhibits a considerable uncertainty in the period, arising from their having adopted different systems of elliptic elements of the orbit of the comet at its former appearance. The time of perihelion passage obtained by the astronomers above named, are as follows:

Lubbock	.	.	.	.	October	31, 1835.
Damoiseau	.	.	.	.	November	4, "
Rosenberger	.	.	.	"	11, "
Pontécoulant	.	.	.	"	14, "
Lehmann	.	.	.	.	"	26, "

The computations of Rosenberger and Lehmann were published only a short time before the comet made its appearance, — those of Lehmann having been announced on the 25th of July, 1835. They

published also an ephemeris of the comet for the period of its visibility, in which its exact route in the heavens was accurately designated, and the date at which it would probably be visible with the aid of a telescope. On the 5th of August, 1835, Dumouchel, Director of the Observatory of the Roman College, at Rome, directed his telescope to the point indicated by Rosenberger's ephemeris; and saw the comet, as a faint and almost invisible stain of light on the deep blue of the heavens, within a degree of its predicted place. On the 20th of August it was generally observed in Europe, and was subsequently followed until about the time of its perihelion passage, on the 16th of November, when it ceased to be visible, on account of its having moved toward the south. It was observed in the southern hemisphere throughout February, March, and April, 1836; and disappeared finally on the 17th of May of that year.

When the comet first became visible, it presented the appearance of a small, round, or somewhat elliptic nebula, without any indications of a tail, and having a point in which the light seemed to be strongly condensed, situated eccentrically within it. On the 2d of October the first indications of a tail were exhibited, which increased rapidly, and on the 5th had attained a length of five degrees. On the 15th of October, nearly a month before its perihelion passage, the tail had attained its maximum length,

18

which was upwards of twenty degrees. From that date it decreased rapidly, and on the 29th of October it was only three degrees in length. It continued to decrease; and on the 16th of November, the date of its perihelion passage, the comet was observed at Pulkova, in Russia, where no tail was visible.

Previous to the return of the comet, although its physical appearance was not expected to be such as to create sensations of terror, even among those inclined to be superstitious, yet the expectation was very general among astronomers that, at its return to its perihelion in 1835, it would afford an opportunity for obtaining new data, from which to derive some satisfactory theory respecting the physical constitution of the class of bodies of which it is so striking an example. It is perhaps unnecessary to state, that it no sooner became visible than phenomena began to be exhibited, preceding and accompanying the gradual formation of the tail, the observation of which has been most justly regarded as forming a memorable era in the history of astronomy, — or more especially in that relating to the cometary bodies. All those strange and important appearances which the comet presented were observed with the greatest zeal, and delineated with the utmost fidelity, by several different astronomers, among whom were Bessel, Schwabe, and Struve, in the northern hemisphere, and Maclear and Herschel at the Cape of Good Hope; each of whom have recorded the successive transformations presented by the comet, under the physical influence of a constantly varying temperature, in its approach to and departure from the sun.

The most striking phenomena which the comet presented early in October, were those which, commencing simultaneously with the growth of the tail, evidently connected themselves with the production of that appendage, and its projection from the head of the comet. On the 2d of October, the nucleus, which had hitherto been faint and small, was observed to have become suddenly brighter, and clearly indicated the commencement of the formation of the tail, by the appearance of being in the act of throwing out a jet or stream of light from that part presented toward the sun. This ejection was by no means either uniform or continuous, but appeared to take place at intervals, like the issuing of the fiery matter from the crater of a volcano. On the 4th of October, as seen through a powerful telescope, the central part of the comet presented a very curious appearance. The nucleus was elliptical, at least five or six times longer in one direction than in the other; and is said to have resembled a burning coal, from which there issued, in a direction nearly opposite to that of the tail, a divergent flame, varying in intensity, in form, and in direction. Occasionally it appeared doubled, and seemed to indicate that a luminous gas was issuing from the nucleus. This ejection, after ceasing for a brief period, was resumed on the 8th of October, and continued, much more violently than before, with occasional interruptions, until the tail ceased to be visible. At this time, Schwabe noticed what he considered to be a secondary train, extending for a short distance in a direction opposite to that of the original tail, and, therefore, toward the sun. Bessel, however, regarded this appearance merely as the renewed ejection of nebulous matter which was afterwards turned back from the sun, as smoke would be by a current of air blowing from the sun in the direction of the real tail. Both the form of this luminous ejection, and the direction in which it issued from the nucleus, were observed to undergo, from time to time, singular and capricious alterations — the different phases so rapidly succeeding each other, that on no two successive evenings were the appearances similar. Sometimes there were two, and often three, nebulous emanations observed to issue from the nucleus in divergent directions. The directions of these emanations of nebulous matter, and also their comparative brightness, appeared to be subject to continual variations. At one time the emitted jet was observed to be single, as above described, and confined within narrow limits of divergence from the nucleus. At other times, it presented a fan-shaped or swallow-tailed form, not unlike that of the flame issuing from a gas-burner with a flattened orifice.

Appearances similar to these were noticed by Arago, at Paris, and also by other astronomers at several different places. On the 15th of October, Arago saw a luminous sector, or diverging light, issuing from the head of the comet, a little to the south of the point immediately opposite the tail, which was much more brilliant than the rest of the nebulosity, and was bounded by two lines of fainter light directed toward the centre of the head. On the following evening no trace of this sector existed; but at a point diametrically opposite the axis of the tail, a new sector was formed, of more than a quarter of a circumference, in angular extent, which was much more elongated and brilliant than the one first seen, and which was bounded by two very bright lines tending to the centre of the head. On the 17th this appearance had diminished in splendor, and on the evening of the 18th — the atmosphere being extremely clear — the entire comet, including the tail, had very sensibly decreased in brilliancy. On the 21st, there were three luminous sectors or brushes of light seen in the nebulosity of the head, the most feeble and dilated

18 *　　　　o

of which was exactly in the prolongation of the tail.
Two days later, these had totally vanished, but the
whole aspect of the comet was completely and
suddenly changed. The nucleus, hitherto so brilliant and well-defined, had become large and diffused; and although it still occupied the centre of
the head, yet the brilliancy of the nebulosity on the
eastern side far surpassed that on the western.

Another remarkable circumstance connected with
the emission of the luminous jets already described,
was that the direction of the principal jet was observed to oscillate to and fro, on either side of a line
drawn from the sun through the centre of the head
of the comet, just as the needle of a compass oscil-
lates on either side of the magnetic meridian. This
vibratory motion was so rapid that the direction of
the jets was observed to be visibly changed from
hour to hour, under the eye of the observer. These
jets, though very bright at their point of emanation
from the nucleus, are described as fading rapidly
away, and becoming diffused as they expanded into
the coma, at the same time curving backwards like
streams of steam or smoke, thrown out more or less
obliquely from narrow orifices, in opposition to a
powerful wind, against which they are unable to
make way, and ultimately yielding to its force, are
drifted back and confounded in a vaporous train,
following the general direction of the current. At

one time, Struve saw the comet thus attended by two delicately-shaped appendages of light, of a most graceful form, one of which preceded, and the other followed the nucleus. Subsequently it appeared to be surrounded by a sort of semicircular veil, which, extending backward, was lost in a double train of light, stretching out to a vast distance from the body of the comet. On the 5th of November, 1835, when seen through the powerful telescope at Pulkova, two flames were seen issuing from the nucleus in nearly opposite directions, both of which were curved toward the same side. The brighter flame, directed toward the north, was marked by strongly defined edges; while the other, directed toward the south, was more feeble and ill-defined. The opposite diagram represents the comet as seen at this time.

The comet was not observed in Europe, with but few exceptions, after the perihelion passage. For upwards of two months it was lost in the sun's rays, but again reappeared on or about the 24th of January, 1836, when it was observed in the southern hemisphere. Its aspect, however, was altogether different from that under which it had been seen previous to its perihelion passage — having, during the interval in which it was invisible, undergone some great physical change, which had operated to produce an entire transformation of its physical appearance. It is described as no longer presenting any vestige of a tail, but as appearing to the naked eye about as bright as a star of the fifth magnitude, shining through a haze; while, with the aid of a powerful telescope, it was seen as a planetary nebula, with a small, round, and well-defined disc, surrounded by a nebulous envelope or coma of considerable extent. Within the well-defined head or disc, and somewhat eccentrically placed, was a vivid nucleus, resembling a miniature comet, with a head and tail of its own, perfectly distinct from and greatly exceeding, in the intensity of its light, the other portions of the comet. A minute bright point, resembling a small star, was distinctly perceived within it, but which was never sufficiently well defined to give any positive assurance of the existence of a solid sphere, nor could any indication of a gibbous form be detected.

The phenomena and changes which the comet presented subsequent to its reappearance in the southern hemisphere, and until its final disappearance, were carefully observed at the Cape of Good Hope. As it receded from the sun, the coma or envelope speedily disappeared, as if absorbed into the disc, which, on the contrary, is said to have increased continually in dimensions, and with such rapidity, that in the week ending February 1st, 1836, the actual volume or *real solid contents* of the illuminated space had expanded at least forty times. This increase of the size of the nucleus, or rather the head, continued with undiminished rapidity until the final disappearance of the comet. The brilliancy of the comet was found to decrease in proportion as the magnitude of the head increased. The nucleus of the comet, properly so called, was not, however, observed to undergo any very considerable change, while the envelope was continually dilating, and becoming fainter and fainter. Moreover, the ray proceeding from it is said to have increased in length and comparative brightness, preserving its direction along the greater diameter of the head, — which had acquired the form of a parabola, — and exhibiting none of those irregular phenomena which characterized the jets emitted previous to its perihelion passage. This ray of nebulous light, however, faded away by degrees; and on the 5th of May the comet appeared precisely as at the time of its first appearance in August, 1835, or as a small round nebula, with a condensation of light near its centre. This was the last observation of the comet in the southern hemisphere; but it was seen twelve days later by Lamont, at Munich, in Bavaria, when it disappeared not to return again until about the end of February, 1911.

The position of the comet in the heavens from day to day was also observed with great precision; and from these observations, combined with those made in 1759, and at previous returns, it will now be possible to investigate all the circumstances of its motion, and to predict the date of its next perihelion passage within a few hours, or, perhaps, within less than an hour. To accomplish this will require immense labor, being sufficient, when carefully performed, to occupy the entire time of one computer for several years. But great as this task may seem to be, it is nevertheless certain that some one will undertake it—if it has not been commenced already —and that the time of the next return of the comet will be so accurately predicted, that the exact place in the heavens to which the telescope must be directed will be announced, and the comet will be seen in strict accordance with the announcement. It is indeed true, that there may be bodies of a planetary or cometary nature in the far distant regions of the heavens in which the comet wanders, which may disturb its motions and change the period of its revolution; but so accurately can the disturbing influence of all the known bodies of our system be calculated, that the existence of these unknown sources of perturbation will be made evident to future astronomers, should any difference be hereafter exhibited between the computed and actual orbit of the comet.

The accuracy of Pontécoulant's prediction of the

perihelion passage of the comet in 1835, affords a
striking instance of the precision of the methods of
calculation employed by him, when we consider the
long period of seventy-five years during which the
comet had been invisible. Moreover, the small discrepancy of only two days can be readily accounted
for by the fact that the existence of the planet Neptune was then hardly suspected, and the masses of
Jupiter and Uranus were not then as well determined as at the present time. And, further, the existence of a medium of resistance has since, as we
shall see, been conclusively established. The influence of this ethereal medium on the motion of this
comet cannot be determined, with great certainty,
until it has performed another revolution. Its effect
on the duration of the last period of the comet has
indeed been computed; but this is a problem which,
in the present state of our knowledge, can be solved
only by experience, since it is evident that the effect
of such a medium of resistance as is here supposed
to exist, must depend upon the magnitude and
density of the comet, and upon the law by which
the density of the ethereal fluid varies with the distance from the sun. It is supposed that this medium
has acquired a rotary motion from east to west —
especially within the limits of the solar system—and
it will therefore be necessary, in order to arrive at
results of the very greatest accuracy, to decide what

part of the variations in the period of its revolution may be due to that cause. These and similar questions are to be carefully discussed before a precise knowledge of the motions of the comet can be obtained; but that all this, and even more, will be accomplished, the history of astronomy during the last fifty years furnishes ample reason to expect.

The influence of the attraction of the planets on the motions of Halley's comet, can be computed with extreme precision; and were there no other disturbances to be experienced during the entire course of its revolution around the sun, it would be possible to predict its future movements with the same precision in which the motions of the planets are determined. It is possible that the orbit of Neptune does not mark the confines of the planetary system, but that there may be other planets still more remote, which, in due course of time, may change the motions of the comet, and which, in case they are not sooner revealed, may finally cause it to fail to appear in accordance with prediction. The time will indeed come, when the action of Neptune will greatly affect the orbit of the comet; and had this planet not been discovered, it would at some future day have falsified the predictions of astronomers, and would, perhaps, have again involved the entire theory of cometary motions in impenetrable mystery. It is with no small degree of pleasure, therefore, that the corresponding advancement of every department of physical science can be contemplated. The discovery of Neptune was itself one of the most brilliant achievements of the human intellect which has been recorded; but it is not until its entire influence in the system is considered, that we can form a true conception of its ultimate importance.

Some idea may be formed of the vast size of the orbit of Halley's comet, by comparing it with the dimensions of that of our earth. The mean distance of the earth from the sun is about 95,000,000 miles, while the length of the orbit of the comet is about thirty-six times this distance, and its breadth about one-fourth of its length. It should be remembered, however, that the eccentricity of the orbit of the earth is very slight, while that of the comet is very great; and, consequently, the distance of the comet from the sun is subject to great variations. In approaching the sun, its velocity continually increases, until it darts around him with astonishing rapidity, at a distance, when in the perihelion, of only 47,000,000 miles from its centre. Its velocity then gradually diminishes, after leaving the sun, till it reaches the most remote point, or aphelion of its orbit, where its distance from him is about 3.373,000,000 miles — a distance which exceeds that of the remotest planet of our system, but at which the attraction of the sun is still sufficient to recall

19

the comet toward him. When, therefore, the comet is in its perihelion, the amount of heat and light which it will receive from the sun, will be four times greater than what is received at the earth. When the comet is in the aphelion of its orbit, the light and heat will be nearly 6000 times less than at the perihelion; and could the sun be viewed from the comet at this enormous distance, it would not appear larger than a star. It is evident, therefore, that the vicissitudes of temperature experienced by the comet must be almost beyond our conception. If the earth could be transported to the aphelion of the comet's orbit, every liquid substance would become solid by congelation, even if the atmospheric air and other permanent gases did not become liquids by the same process. But if, on the other hand, the earth were placed at the perihelion of the orbit of the comet, all the liquids upon it would be converted into vapor, either mixing with the atmospheric air, or arranging themselves in regular strata, one above the other, according to their relative weights.

We have thus described, somewhat in detail, the most important events connected with the discovery of the periodic character of Halley's comet, and also the most interesting phenomena which it presented. We have seen that it exhibited to a high degree those physical phenomena which seem to belong almost exclusively to this class of bodies, and that it affords an unparalleled example of the result of those powers of calculation by which we are enabled to follow it, in the mind, through the depths of space, far beyond the extreme limits of our system, and, notwithstanding disturbances which render each succeeding period of its return different from the last, to foretell that return with the utmost precision.

The next comet, whose period of revolution has been definitely determined, which we shall consider, is that known as Encke's comet, and which, although one of the most interesting in its phenomena, is still, in almost every respect, unlike that which we have just described. This comet was discovered by Pons, at Marseilles, on the 26th of November, 1818. It was not visible to the naked eye; and when seen through a telescope, presented the appearance of an ill-defined nebulosity, without a tail, but evidently having a point in which the light was strongly condensed, situated eccentrically within it. As soon as it had been observed to a sufficient extent, the elements of its orbit were computed; and on comparison of the elements of those comets which had previously appeared, it was immediately noticed that this comet was identical with one which was visible in 1805. It was uncertain, however, whether the period was thirteen years, or whether the comet had, in the mean time, returned unperceived to its perihelion. Subsequently, Encke undertook to remove this doubt, by computing an elliptic orbit directly from the observations made in 1818 and 1819, during the entire period in which the comet was visible; and he found that its period of revolution was only about 1200 days, and that, consequently, the comet had returned three times between the years 1805 and 1819. It was found, also, upon further examination, that the same comet had been observed in 1786 and in 1795. Having thus identified the comet at four different returns to its perihelion, Encke was enabled to ascertain the period of its revolution with great precision; the result being 1208 days. This, however, occasioned great astonishment, and some degree of doubt, since it was generally believed that the periods of comets must necessarily be very long.

The identification of the comet with those of 1786, 1795, and 1805, enabled Encke to use observations including an interval of nearly thirty-three years in making his computations; and having computed its motion from the year 1819, he predicted that, as its period would be reduced to 1203 days by the disturbing action of the planets, it might be expected to appear about the beginning of June, 1822, but that, on account of its position in the heavens, it would only be visible in the southern hemisphere. The return of the comet was therefore looked for by the astronomers who happened to be in that part of the globe; and on the 2d of June,

1822, it was actually discovered by Rumker, at Para-
mata, in New South Wales, and observed until the
end of the same month. The next return was predicted to take place in 1825; and on the 13th of
July—true to its appointed course—the comet was
observed by Valz, at Nismes. The next return took
place in 1828. It was first seen by Struve, at Dor-
pat, in Russia, on the 13th of October of that year,
and was observed in the European observatories till
December 25th.

In 1832, the comet again returned to its perihelion;
but being unfavorably situated for observation in
the northern hemisphere, it was only seen by Hard-
ing, at Göttingen, on the 21st of August. It was
observed, however, by Henderson, at the Cape of
Good Hope, during the entire month of June, and
was also seen at Buenos Ayres. In 1835 it was observed from July 22d till August 6th, and in 1838 it
was seen at Breslau, on the 14th of August, as a
very faint, ill-defined object. It subsequently increased in brilliancy, and continued visible until the
middle of December. In combining all the observations which have been made from 1786 to 1838, inclusive, Encke found it necessary to change the
value which he had hitherto employed for the mass
of the planet Mercury, in computing the perturbations of the comet, and also to make allowance for
the supposed influence of a resisting ethereal fluid,

19 *

pervading the regions of space, as already alluded to in the case of Halley's comet. It should be remarked, however, that even before this, the comet had indicated that a correction to the adopted value of the mass of Jupiter was necessary. The plane of the orbit of the comet being inclined to that of the orbit of Jupiter by a small angle, it necessarily approaches very near the planet, and, consequently, is subject to considerable disturbance, in its motions, from this source. Under these circumstances it may be perceived that, since the action of Jupiter on the comet depends on their distance and relative quantities of matter, any error in the assumed value of the mass of the former would be exhibited by the excess or deficiency of the resulting values of the perturbations. But Laplace had computed, from the theory of probabilities, that there was hardly any chance that the mass of the planet, which was generally used, could be in error even by the one-hundredth part of its value.

The discrepancies, however, which were plainly exhibited in the motions of the comet, were such as to excite a strong suspicion of its inaccuracy; and accordingly a new determination of this important quantity was attempted. The result was that it was found by three different computers, each by an entirely different process, that the mass of Jupiter, hitherto assumed, was in error more than four times the amount which Laplace had assigned as the limits of its probable uncertainty. With this new value of the mass of Jupiter, and admitting the existence of a resisting medium, it seemed possible to announce the future movements of the comet with great precision. The next return was therefore predicted to take place in 1842, and on the 8th of February of that year the comet was detected by Galle, at Berlin. It was observed in Europe and in the United States until near the middle of April, and was found to follow the computed path very closely. In 1848 it was observed for a period of three months, extending from August to November. When first detected it was very faint, but towards November it had increased in brilliancy so as to be barely perceptible to the naked eye as a dull hazy star of the sixth magnitude. The observations made at this return of the comet have not only confirmed the supposed existence of an ethereal medium of resistance, but have also been the means of deriving some very important results in relation to the changes which its magnitude or volume underwent in approaching the sun, and also in receding from it.

The next return of the comet, which has received the name of *Encke's Comet*, in honor of the discoverer of its periodic character, took place in 1852. It passed its perihelion on the 14th of March of that year, but was first seen on the 2d of January. The

next return occurred in 1855; but the comet was
not favorably situated for observation in the northern hemisphere. It was observed at the Cape of
Good Hope, and also in Santiago, in Chili. The
comet again appeared in 1858, almost exactly in the
place assigned by computation. It came to its
perihelion on the 18th of October of that year; but
was first observed, at Berlin, on the 7th of August.
From the observations made at the Cape of Good
Hope in 1855, in combination with those made previously, at the successive returns of the comet, Professor Encke determined its orbit anew; and having
made due allowance for the planetary perturbations,
he constructed an ephemeris, and on the first clear
evening after the comet was computed to be in a
position for observation, the telescope was directed
to the spot indicated, and the comet was barely distinguishable as a dim nebulosity, without any distinct or regular outline. In a few weeks, however,
the brilliancy had increased to such an extent, that
when seen through a telescope, the comet appeared
as a round, well-defined nebula, without a tail, but
having decided indications of a point near the centre,
in which the light was strongly condensed. About
the middle of September it became visible to the
naked eye, but only as a star of the sixth or seventh
magnitude, shining through a thin haze. At this
time the comet was seen in the east several hours

before sunrise; and being in a position very favorable for observation, it was possible to detect a very slight elongation of the head of the comet in a direction opposite the sun. The nucleus was almost a stellar point, with a coma or envelope of very dense nebulous matter. The comet was not observed after the first week in October, having been lost in the approaching twilight. The next return of the comet will take place in March, 1862.

The orbit of Encke's comet is an ellipse of great eccentricity, whose length is about double its breadth. When at its perihelion it is distant from the sun only 32,000,000 miles, or a little less than the distance of Mercury. When in the aphelion of its orbit the comet is distant from the sun about 400,000,000 miles, or a little more than four times the mean distance of the earth. One extremity of the orbit, therefore, reaches a little beyond the orbit of the asteroid Pallas, and the other extends to that of Mercury. The length of the orbit, however, is found to be gradually decreasing; and, consequently, the period of revolution is becoming shorter and shorter. This phenomenon has been attributed to the existence of a resisting medium, consisting of a subtle ethereal fluid, pervading all space. The decrease of the period of revolution of the comet, after making due allowance for the perturbations which have been caused by the action of the planets,

P

is exhibited in the following table, from 1786, the date of its first appearance, till 1858, the time of its last return to its perihelion:

Date.	Period of revolu-tion in days.	Date.	Period of revolu-tion in days.
1789,	1212.79	1825,	1211.55
1792,	1212.67	1829,	1211.44
1795,	1212.55	1832,	1211.32
1799,	1212.44	1835,	1211.22
1802,	1212.33	1838,	1211.11
1805,	1212.22	1842,	1210.98
1809,	1212.10	1845,	1210.88
1812,	1212.00	1848,	1210.77
1815,	1211.89	1852,	1210.65
1819,	1211.78	1855,	1210.55
1822,	1211.66	1858,	1210.44

The existence of some retarding force is here very strongly exhibited. The action of the planets has been computed with the greatest possible precision; and yet the period is decreasing, by slow and regular degrees, at the rate of about eleven hundredths of a day, or about two hours and thirty-eight minutes, at each successive revolution. This is evidently the result which would be produced by a resisting medium, such as is supposed to exist, since the effect of the resistance is to diminish the velocity of the comet in its orbit, thus lessening the force resulting from the orbital motion, or what is called the centrifugal force, and consequently compelling the comet to fall towards the sun, in obedience to the now preponderant force of gravity, until the equilibrium is again restored. The comet will therefore be continually describing a new orbit,—smaller than the one in which it had previously moved,—and with an increased velocity. Thus it may be perceived that the medium of resistance actually accelerates the motions of the comet; but since the force of resistance is confined to the plane of the orbit, it can have no influence whatever in producing any additional changes in the elements. In computing the effect of the resisting medium, it is assumed to increase inversely as the squares of the distances, and that its resistance acts as a force tangent to the orbit, and proportional to the squares of the comet's actual velocity in each point of its orbit. We shall subsequently give a complete explanation of the theory of this ethereal medium, and also consider its ultimate effect on the motions of the heavenly bodies, and more especially on those of the comets.

This comet, as we have already stated, is a small telescopic object, of extreme tenuity, very pale and diffused in its appearance, even at the point of greatest condensation, and without a tail. It has, however, although apparently an insignificant object, opened a wider field of discovery, and more new and interesting facts, than any other on record. It has, as above noticed, furnished the means of proving the existence of an ethereal fluid, pervading

all space; and has afforded a striking instance of
the changes which take place in the nebulous matter
on approaching the sun, and in retreating from him;
and also, an opportunity of measuring these changes.
Stars which have been seen through it more than
once, have attested the high degree of its attenuation; while the smallness of its orbit, and the short-
ness of its period, give it almost the characteristics
of a planet. The constant decrease in the size of
its orbit, arising from the resistance of the ethereal
fluid, may possibly precipitate it upon the surface
of the sun, if, before that event can take place, the
perpetual diminution of its mass does not offer an
example of condensation to the solid state, or of
annihilation; and its perturbation by the planets
have been the means of indicating errors in the
assumed masses of Mercury and Jupiter.

At each successive return, this comet furnishes
additional evidence in support of the hypothesis of
a resisting medium. Indeed, it has already incon-
trovertibly established the fact of the existence of
such a medium, which, if there be not other considerations to counterbalance, may result ultimately,
though at a time almost infinitely distant, in the
total annihilation of the planetary system. It is a
result so startling in its consequences, and so repug-
nant to our preconceived ideas of the stability of
our system, that the mind almost shrinks in horror

from its contemplation. If it were not that the calculations from which such a conclusion has been
derived, have been repeated by different individuals,
each one using different methods of computation,—
but all arriving at the same general results,—we
might perhaps hesitate to receive it. But when evidence of the most unequivocal character has been
adduced, we are compelled to admit the fact, and
look to other forces to prevent the inevitable consequences which must ensue. There can be but little
doubt that there are other forces in nature than
those now known, which time alone will unfold; and
which will enable the astronomers of some future
day to explain all these mysterious questions, to
dispel all such dismal forebodings as the existence
of a resisting medium in space, with our present
knowledge, will excite; and to furnish finally the
most sublime proofs of the fixed design of perpetuity which the Almighty Creator has enstamped on
his universe of creation. Let us, therefore, be content with our present knowledge of laws and forces;
and, if it be not permitted to our day and generation, leave it to those who shall come after us to
solve these difficult problems.

The discovery of this comet's periodic character was
soon followed by the announcement of a second comet
of short period. This comet was detected by Biela,
an Austrian officer, at Josephstadt, in Bohemia, on

20

the 27th of February, 1826; and was subsequently
independently discovered by Gambart, at Marseilles,
on the 9th of March of the same year. It was a
faint, nebulous object, not unlike Encke's comet in
general appearance, and was observed only about
two months. As soon, however, as the elements of
its orbit had been computed, it was found that the
comet was probably identical with one which had
appeared in 1805, and for which Gauss had obtained
an elliptic orbit, corresponding to a period of about
six years. A more complete investigation of the
orbit showed that it arrived at its perihelion a little
before midnight, at Paris, on the 18th of March,
1826; and that the orbit in which it was then moving
was an ellipse, giving a period of revolution of 2460
days, or about six years and nine months. It was
also found, in tracing its path backwards, that the
comet was observed in 1772 and in 1805; and that,
between the former date and 1826, it had performed
eight revolutions. It was therefore predicted that it
would return again in 1832; and on the 26th of
November, of that year, it passed its perihelion in
almost exact accordance with the prediction. It was
observed during the months of October, November,
and December; and was last seen, at the Cape of
Good Hope, on the 3d of January, 1833. At this
return it was very faint, and was observed with the
greatest difficulty.

BIELA'S COMET, 1846.

TUTTLE'S COMET.

The orbit of this comet is an ellipse whose eccentricity is a little less than that of Encke's comet, or about 0.75. Its perihelion distance is nearly 84,000,000 miles, and its aphelion distance 581,000,000 miles. It therefore comes within the orbit of the earth, and recedes beyond the orbit of Jupiter; and since the plane of its orbit is inclined to the plane of the ecliptic by a small angle, it is evident that the comet may approach very near these planets, and thereby be subject to considerable disturbance in its motions. Moreover, it happens that one of the points in which Biela's comet intersects the plane of the ecliptic, is at a distance from the earth's orbit less than the sum of the semi-diameters of the earth and comet. It follows, therefore, that in case the earth should ever arrive at this point of its orbit at the same moment at which the comet passes through the point of its orbit which is nearest to it, a portion of the head of the latter would envelop the earth. It was, indeed, found, on the occasion of the return of this comet in 1832, that if, by any unforeseen circumstance, the comet should be delayed one month, it would come in contact with the earth on the 30th of November. This announcement created a profound sensation, and no small degree of alarm among those unacquainted with the character and motions of such bodies. But these apprehensions were wholly unfounded — the comet having passed that point of its orbit on the 29th of October, at a distance from the earth of more than 60,000,000 miles.

At its return in 1839, the comet was most unfavorably situated for observation, being in the immediate vicinity of the sun; and consequently is not known to have been seen from any part of the globe. The next return, however, occurred under very favorable circumstances. From the observations made at its previous returns to its perihelion, Santini determined the elements of its orbit with great care; and having computed the perturbations of the planets during the period in which it was invisible between 1832 and 1845, he computed an ephemeris, by means of which it was discovered by De Vico, at Rome, on the 26th of November, 1845. It was seen at Berlin on the 28th of November; at Cambridge, England, on the 1st of December; and subsequently at all the principal observatories in Europe. It continued to be visible until the end of April, 1846, having passed its perihelion on the 11th of February of that year.

This comet, as already stated, is a small, round, nebulous object, not visible to the naked eye even when brightest, and without a tail. It presents no solid, or even well-defined nucleus; and appears to be a mere mass of vapor, so extremely attenuated that minute stars can be seen through it, although its diameter probably exceeds 50,000 miles. The deficiency, however, in brilliancy and magnitude, is fully compensated by the anomalous phenomena which it presents, of two distinct comets, each complete in itself, and revolving around the sun in such close proximity as to be almost always visible simultaneously in the field of the telescope. In 1832, and at its previous returns, it appeared as a single body, with a point of condensed light near its centre; and when first observed in 1845, there was nothing unusual noticed in its appearance. About the middle of December, 1845, it was noticed that the comet gave positive indications that some violent change of its form was about to be effected. On the 19th of December, it is described as being pear-shaped; the nebulosity being unduly elongated in a direction different from that in which a tail would be seen. On the 29th of December, it was seen attended by a faint, nebulous spot, which gave it the appearance of having a short tail connected with the envelope of the nucleus by a denser mass of nebulous matter. But on the 13th of January it was noticed at Washington, that instead of being, as usual, a single comet, it apparently consisted of two comets, moving through space side by side. Each body is said to have exhibited all the characteristics of a telescopic comet, being gradually condensed towards the centre, without any distinct and

20 *

well-defined disc; and each elongated on the side
opposite the sun.   They were, however, of unequal
size; one being at least eight times larger than the
other, and about four times brighter.

The separation of the comet into two distinct
parts, was observed at several places in Europe, on
the 15th of January, 1846, and, subsequently, at all
the principal observatories.   The smaller comet was
supposed to have been thrown off from the nucleus
of the larger one, which was regarded as the original
comet.   The latter was, as above noticed, at first a
globular mass of nebulous matter, semi-transparent
at its very centre, and without any tail; while, after
the separation, both comets showed decided indications of short trains, parallel in their direction, and
at right angles to the line joining the centres of the
nuclei.   Again, it was observed that from the day
of their separation, the original comet decreased
both in magnitude and brilliancy, and the companion
increased until near the middle of February, when
they were sensibly equal.   After this the companion
still increased, and from the 14th to the 16th of
February, was not only much brighter than the
original comet, but had a distinct stellar nucleus.
The change of brilliancy was again reversed, the
original comet recovering its superiority in this respect, and acquiring, on the 18th of February, the
same appearance which was presented by the com-

panion from the 14th to the 16th of the same month.
After this the companion gradually faded away, and
on the 1st of March could only be seen with great ·
difficulty. It was last observed on the 15th of
March, and subsequently the comet again appeared
to be single.

While these anomalous changes were taking
place, there were indications of a connection between
the two nuclei, a faint arc of light being noticed,
which extended as a kind of bridge from the one to
the other. This line of light, however, did not
occupy any fixed position, but was observed to have
at times a slight angular motion, sometimes in one
direction and sometimes in the opposite direction,
and thus the comets were seen to oscillate about
each other, according to some mysterious law which
remains yet to be fully revealed.

The opposite diagram represents the appearance
of the comet as seen through a telescope about the
middle of February, 1846.

The orbits of both the original comet and its companion, have been computed as if they were separate
and independent bodies, by Plantamour, of Genoa;
and he finds that the real distance between their
centres was, during the greater portion of the period
of their visibility, subject only to very slight variations, being about 150,000 miles. The same computations were performed by Hubbard, of the Na-

tional Observatory, at Washington, who finds that
the entire series of observations may be satisfactorily
represented by supposing each nucleus to describe
an independent ellipse around the sun. It is found
also that the comets moved on side by side, without
manifesting any reciprocal disturbing action; or
that, at least, if any such disturbance did exist, it
was too minute to be indicated by the observations
—a circumstance which shows conclusively that the
masses of such bodies are almost infinitely small.
The following table gives the distance of one nucleus
from the other, as determined by means of Hubbard's elements of the orbits:

Date.	Distance, in miles.	Date.	Distance, in miles.
1845, Dec. 1,	. . . 117,000	1846. Feb. 17,	. . 226,000
1846, Jan. 20,	. . . 198,000	" " 25,	. . 216,000
" " 28,	. . . 213,000	" March 5,	. . 202,000
" Feb. 9,	. . . 227,000	" " 21,	. . 170,000

The next return of Biela's comet was predicted
to take place in 1852; and on the morning of the
26th of August, of that year, in searching for the
comet, Professor Secchi, at Rome, detected a small
nebulous object, which was very far from the position
of the comet given by the ephemeris. It was soon
found that the faint nebulosity observed was really
a comet, and still later that it was the one sought
for. No companion, however, was visible; but on
the morning of the 16th of September following,

this was also detected. It was very faint, and of an
elongated ovoid form, the apex being turned away
from the sun. It was distant from the other part—
which was supposed to be the original comet, in
1846—a little more than half a degree. The latter,
at the same time, appeared quite irregular, and had
two very faint streaks of light issuing from it, but
did not exhibit a distinct nucleus, although more
luminous in the centre than at the edges. The
original comet was last seen on the 25th of September, and the companion on the 28th of the same
month. The distances of the two comets from each
other were as follows :

Date.	Distance, in miles.	Date.	Distance, in miles.
1852, Aug. 27,	. . 1,501,000	1852, Sept. 16,	. . 1,616,000
" " 31,	. . 1,531,000	" " 20,	. . 1,624,000
" Sept. 4,	. . 1,559,000	" " 24,	. . 1,625,000
" " 8,	. . 1,582,000	" " 28,	. . 1,619,000
" " 12,	. . 1,601,000		

It is evident from this table of the distances of
the nuclei from each other, that in 1852 the two
comets moved nearly as in 1846, with the exception
only that, at this return, the distance of the two
bodies was nearly seven times greater than in 1846.
We might expect, therefore, that it would be easy
to determine the orbit of each nucleus with great
precision, from the observations made in 1846 and
in 1852. But it is found, unfortunately, that it is

nearly impossible to decide which of the two comets observed in 1852 is the original comet of 1846, since all the observations are represented almost equally well, by either the hypothesis that the one discovered by Secchi, on the 25th of August, was the original comet of 1846, or that the one found on the 16th of September is the identical one. The first supposition has been most generally adopted, and the time of the next return of the comet, in 1859, was computed in accordance with this assumption. It was predicted that the original comet would arrive at its perihelion on the 24th of May, 1859, a little before midnight at Greenwich, and that the companion would arrive at the point of its orbit nearest the sun about thirty-five hours earlier. They were, however, situated in a position very unfavorable for observation, being in the immediate vicinity of the sun; and although there can be no doubt but that they returned very nearly in accordance with the prediction, yet for the reasons here stated, no observation is known to have been made. This is much to be regretted, since a few observations in 1859 would have definitely decided the question of identity of the two comets at the previous returns, and would have greatly facilitated all future investigations respecting their motions. The comets will again return in 1866, and will then be in a position favorable for observation; and we must, therefore, await this appearance in order to settle the important question of identity.

Such are the most prominent characteristics of the phenomena presented by Biela's comet, at its successive returns to its perihelion ; and on account of the anomalous appearance of a double comet here exhibited, it may not be improper to introduce, in this connection, a statement of the various hypotheses which have been devised, in order to explain such an unexampled departure from the general form of such bodies. In the first place, then, it is found, by comparing the distances of the comets from each other in 1846 and in 1852, that the separation must have taken place about 500 days before the perihelion passage in 1846, or about the middle of July, 1843. This date, however, is still quite uncertain; and it will not be possible, until this has been definitely determined, to decide under what circumstances the separation of the nucleus occurred, and by what means it was effected in such a way as not to change or influence the motions of the primary nucleus. Again, the extraordinary changes in brilliancy which the two comets exhibited, both in 1846 and 1852, clearly indicate that the brightness of these objects does not depend merely on their distance from the earth and sun, but upon other unknown causes. These causes may, indeed, have been sufficient to develop the

brilliancy of the companion of Biela's comet, at its
last two returns to the sun, to such an extent as to
render it visible from the earth; while, at its former
returns, on account of its unfavorable position, and
the inefficient operation of these causes, it might
have been too faint to be noticed. The great change,
however, in the distances of the nuclei from each
other from 1846 to 1852, seems to be sufficient in
itself to show that the separation did not take place
until after the return of the comet in 1832.

The most plausible view of the cause of this permanent separation of a comet into two distinct parts,
is that which assumes a repulsive force emanating
from the sun, the nature and effect of which we
shall fully consider when we come to treat of the
theory of the physical constitution of comets, and
of the formation of their tails; it being only necessary to state here, that the phenomena presented by
Halley's comet in 1835, and by the great comet of
1858, have demonstrated the existence of such a
force. The effect of a repulsive force upon the
nebulous envelope of the comet, would be to distort
it from the spherical form,—which it would assume
if the force of gravity alone acted,—causing it to
become an ellipse, the length of which, compared
with its breadth, would be greater as the repulsive
force increases. The repulsive force may now be
conceived to become so great as to drive the remoter

particles of the envelope beyond the influence of the nucleus, and carry them off into space. The entire mass of the comet will therefore be urged toward the sun by the difference of the total attractive and repulsive forces. So long as the repulsive is insufficient to separate the matter composing the comet, it will revolve as a single body, elongated in a direction opposite the sun; but if the repulsive force should be so great as to overcome the gravitation of the nucleus, the exterior portions of the envelope will be successively driven off into space, until finally a new comet is formed, which will henceforth revolve as a companion to the original one, as in the case of Biela's comet.

The successive returns of this comet have also indicated the existence of a resisting medium, the period being always less than at the preceding perihelion passage. In 1852, it came to the perihelion nearly two days in advance of the predicted time, after due allowance had been made for the action of the planets on the comet during the entire interval elapsed since 1846.

On the 22d of November, 1843, a telescopic comet was discovered by M. Faye, of the Royal Observatory, at Paris, which was subsequently found to move in an elliptic orbit, with a period of revolution of about seven and a half years. When first discovered, the comet had a brilliant nucleus, and a short

21   Q

fan-shaped tail, about one-tenth of a degree in length. It was last seen at Pulkova, in Russia, on the 10th of April, 1844. It passed its perihelion on the 17th of October, 1844, at a distance from the sun of 161,000,000 miles. Its aphelion distance is 563,000,000 miles. The comet, therefore, does not come within the orbit of the planet Mars, and recedes, in its aphelion, beyond the orbit of Jupiter. But the most remarkable feature connected with the orbit of this comet, is the eccentricity, which is only a little more than that of the orbits of some of the asteroid planets, thus forming a connecting link between the cometary and planetary systems.

As soon as the entire series of observations of the comet, made in 1843 and 1844, had been published, a complete investigation of the elements of its orbit was accomplished; and the result was, that its period of revolution was found to be 2718 days, or a little less than seven and a half years. It was also noticed that the position of its orbit is such that the comet is liable to excessive perturbations, on account of its close proximity, when near the aphelion, to the orbit of Jupiter. In 1840 it was so near this planet, that his attraction was about one-tenth part of that of the sun, which must necessarily have caused a great alteration of the form of its orbit. In 1815, also, it probably came still nearer to Jupiter, and the perturbations must have been much greater than in 1840. Under such circumstances, M. Valz was led to conclude that, prior to 1840, the comet moved in an orbit entirely different from that in which it is now moving; and that, possibly, this comet might be identical with a comet which appeared in 1770, and which, although at that time moving in an ellipse of only five years, has, by the action of Jupiter, been so excessively disturbed in its motions that it has not since been seen.

A complete investigation of this problem was subsequently undertaken by Le Verrier, who finds, after computing the perturbations of Faye's comet, by Jupiter, that it cannot be identical with the comet of 1770. He also computed the perturbations of the comet arising from the attraction of the planets during the interval from 1843 to 1851, and predicted that it would return to its perihelion on the 3d of April, 1851. With the aid of Le Verrier's elements, an ephemeris was computed for the latter part of 1850, and beginning of 1851, by means of which observers might be enabled more easily to detect the comet at its return; and more especially because it was foreseen that it would be extremely faint and small, and not capable of being seen except through the most powerful telescopes. By means of this ephemeris the comet was re-discovered at the observatory at Cambridge, England, on the 25th of December, 1850, and was followed until the

4th of March, 1851. It was subsequently observed at Cambridge, Mass., and at Pulkova, in Russia; and the positions actually observed scarcely differed from those predicted by means of Le Verrier's elements. The comet was described as an extremely faint object; so much so as to be barely visible through the large telescopes of those observatories. It appeared to be slightly elongated in the direction of the sun, but no well-defined nucleus or point of condensation could be perceived. It came to its perihelion on the morning of the 2d of April, 1851, or within one day of the predicted time. This difference is undoubtedly due, to a certain extent, to the influence of the resisting medium.

The next return of this comet took place in 1858. It was observed during the months of September and October, and passed its perihelion on the 12th of the former month. After making due allowance for the action of the planets from 1851 to 1858, it is found that Le Verrier's elements still represent the motion of the comet with great precision. At its last return, in 1858, the comet was very faint and ill-defined, and without either a nucleus or a tail, being, apparently, a diffused mass of nebulous matter. The next return to the perihelion will occur in March, 1866.

A fourth comet of short period was discovered by De Vico, at Rome, on the 22d of August, 1844,

which, in the moonlight, was almost visible to the
naked eye. It was also discovered independently
by Encke, at Berlin, on the 5th of September, and
on the following evening it was seen at Hamburgh,
by M. Melhop, an amateur astronomer. In America
it was first seen by Mr. H. L. Smith, at Cleveland,
Ohio, on the 10th of September. Soon after the
middle of this month it became distinctly visible to
the naked eye, and when seen through a telescope,
the nucleus presented a very beautiful appearance,
being bright and strongly condensed. The tail was
about one degree in length, extending in a direction
opposite to the sun. It subsequently decreased
rapidly in brilliancy, and was observed for the last
time at Pulkova, on the 31st of December. The
orbit was computed, as usual, on the supposition that
it was a parabola—since this is more easily computed
than an elliptic orbit—but it was soon found that it
was impossible to represent the observations by such
an orbit, and accordingly an ellipse was computed,
which gave a periodic time of about five and a half
years. It passed its perihelion on the 2d of September, 1844, at a distance from the sun of about
114,000,000 miles. When in the aphelion of its
orbit, its distance from the sun is 475,000,000 miles.
The inclination of the plane of its orbit to the plane
of the ecliptic, is less than three degrees; and, consequently, it may come so near the planet Jupiter

21 *

that its orbit will eventually be entirely changed by
the perturbations caused by this planet. It also
comes very near the earth, and is considerably disturbed by its action.

The interest which the appearance of the comet,
in connection with its short period, it being less than
that of any other which had hitherto been discovered, with the exception of Encke's comet, induced the Academy of Sciences, at Amsterdam, to
make the investigation of its orbit the subject of a
prize. Accordingly, a thorough determination of
the elements of its orbit from the entire series of
observations, embracing a period of more than four
months, was undertaken by Brünnow, at that time
assistant in the Royal Observatory at Berlin. He
computed the perturbations of the comet by all the
planets within the orbit of Uranus, excepting, of
course, the asteroid planets between Mars and Jupiter, and finally obtained an orbit which satisfied all
the observations with extreme precision. The period
of revolution was thus found to be $1996\frac{1}{2}$ days, or
very nearly $5\frac{1}{2}$ years. The comet would, therefore,
return again to its perihelion about the middle of
February, 1850; but it happened, very unfortunately,
that it was so near the sun during the entire period
in which it remained sufficiently near the earth to
be visible, if seen under more favorable circum-

stances, with respect to its apparent position in the heavens, that it could not be observed.

The next return of the comet to its perihelion occurred in the summer of 1855, and since it would then be in a position favorable for observation, its reappearance was contemplated with no small degree of interest. The perturbations were computed from 1844 to 1855, and Dr. Brünnow announced that it would be in its perihelion on the evening of the 6th of August, 1855. As soon as the comet was computed to be near enough to be visible from the earth, it was sought for in the position indicated by the ephemeris, but without success. The search was continued until the comet was probably out of reach of the most powerful instruments, and it is not certainly known that it was seen at any part of the globe. On the 16th of May, 1855, Goldschmidt, of Paris, while searching for the comet, found a faint nebulous body, with an ill-defined and irregular outline, and without a tail, in a position which did not differ much from the computed place of De Vico's comet, if it be assumed that the perihelion passage took place on the 11th of August. The comet, however, must have been much nearer the earth on the 1st of August than at the date of this observation, while its position in the heavens must have been much more favorable for observation. It seems certain, therefore, that, since the search continued until after the beginning of August, this nebulous object could not have been the comet; and we are compelled to conclude that it was nowhere seen in 1855—a failure to realize the predictions, which may be accounted for, not by supposing the methods of computation to have been inaccurate, but rather by the uncertainty of the ephemeris, owing to the great difficulty of determining the time of perihelion passage in 1855 from the observations made in 1844. The comet may be expected to return again toward the latter part of 1860, or in the beginning of 1861; but it will not probably be in a position favorable for observation. This is to be regretted, since the comet, before it has accomplished the succeeding revolution, will be subject to great perturbations by Jupiter, which may so completely change its orbit, that it will be difficult to identify it at any future return.

Another important circumstance to be noticed in connection with this comet, is that its orbit is very similar to that of a comet which appeared in 1585. Some astronomers have, for this reason, supposed that De Vico's comet is identical with the comet of 1585; and possibly, also, with those of 1743, 1766, and 1819. Le Verrier and Brünnow have undertaken to decide this question by direct computation; and they conclude, finally, that this comet cannot be identical with either of those just mentioned, nor

with any other on record, unless it be with one
which appeared in 1678. They find that there is a
strong probability of identity in this case, although
the perturbations in the meantime have very materially changed the elements of its orbit.

Another periodic comet was discovered by Brorsen, at Kiel, in Denmark, on the 26th of February,
1846. It was a faint, telescopic object, and was
found to move in an elliptic orbit, with a periodic
time of about five and a half years; and has been
supposed to be identical with the comets of 1532
and 1661. It was observed until the 22d of April;
and from the entire series of observations, its exact
period was ascertained to be 2039 days. Its perihelion distance is about 62,000,000 miles. When in
the aphelion, its distance from the sun is 532,000,000
miles. The next return to the perihelion took place
in September or October, 1851; but it was in a position very unfavorable for observation, and consequently was not seen. Moreover, the small number
of observations which were made in 1846, rendered
it extremely difficult to determine the precise period
of its return, otherwise it would probably have been
observed in 1851. The next and last return took
place in 1857; but the comet was accidentally discovered by Bruhns, at Berlin, on the 18th of March,
and its identity established only after parabolic elements had been computed which were very similar

to those of Brorsen's comet. The supposed identity which resulted from these approximate elements, was subsequently fully confirmed; and the comet is now ranked among those of short period whose future movements can be foretold with the utmost precision. It passed its perihelion on the 29th of March, 1857; and may be expected to return again in October, 1862.

A sixth comet of short period was discovered by D'Arrest, at Leipsic, on the 27th of June, 1851. It was a faint, telescopic object, without a tail, but having a distinct and well-defined outline, with a point of greater condensation near its centre. It passed its perihelion on the 8th of July, 1851; and was found to move in an elliptic orbit corresponding to a period of revolution of 2353 days, or nearly six and a half years. The next return was therefore predicted to take place about the end of 1857, or beginning of 1858; and the comet was observed at the Cape of Good Hope in December, 1857, and in January, 1858. The next return will occur in June, 1864.

The seventh comet of short period, whose orbit is definitely known, was discovered in the constellation Andromeda, by Mr. Horace P. Tuttle, at the observatory of Harvard College, Cambridge, Mass., on the evening of the 4th of January, 1858; and was subsequently independently discovered by Bruhns,

at Berlin, on the 11th of the same month. At
the time of its discovery it appeared like a faint and
diffused nebula, without any indications either of a
tail, or of an elongation in a direction opposite to
the sun. At the beginning of February it had increased very considerably in brilliancy, and presented the appearance of a well-defined elliptical
nebula, about three minutes of an arc in diameter,
with a distinct increase of light from the borders
toward the centre. It continued visible until near
the end of March — the last observation reported
from any part of the world being that made at
Cambridge, Mass., on the 23d of this month. When
last observed, although then very far south, and
apparently enveloped in the vapors near the horizon,
it nevertheless appeared distinct, and nearly as brilliant as when it was first discovered; so that, in
case it had been detected at some observatory in the
southern hemisphere, it might have been followed
at least one month longer. The observations, however, which were made in Europe and in the United
States during the period of its visibility in the northern hemisphere, have enabled astronomers to determine the elements of its orbit with considerable
precision.

The first approximate elements of the orbit of this
comet were found to bear a striking resemblance to
those of the second comet of 1790, which was dis-

covered by Mechain, at Paris, on the 9th of January of that year. Indeed, so close was this resemblance, that the identity of these two comets was placed beyond a doubt; but it still remained to determine whether the comet had not previously returned to its perihelion unperceived, during the interval of sixty-eight years which had elapsed between 1790 and 1858. It was found by direct calculation of elliptic elements, that the period of revolution is about thirteen and a half years; and, consequently, that the comet had returned to the perihelion no less than five times between 1790 and 1858. It might be expected, however, that since the orbit of the comet is so situated that the perturbations can never be very considerable, the comet would have been observed at one or more of its intermediate returns. In the latter part of 1830, and in the beginning of 1831, the comet was favorably situated for observation, both in respect to its apparent angular distance from the sun, and its brilliancy. It might at this time have been easily observed in the east before sunrise; and it must be supposed, therefore, that no search was made for comets in that portion of the heavens in December, 1830, and in January, 1831. At the other three returns of the comet to its perihelion, it was in such close proximity to the sun, and — on account of its greater distance from the earth — so diminished in brilliancy, as not to come within the range of the optical power of such instruments as were usually employed in searching for cometary bodies. It might likewise be supposed that the comet would have been observed at some return to its perihelion previous to 1790; but it should be remembered that, until near the commencement of the present century, no systematic search for faint, telescopic comets, had been instituted; and that its discovery at a date prior to 1790, could only have been the result of a very fortunate accident.

The opposite diagram represents the appearance of this comet on the 31st of January, 1858, and may be regarded as a general delineation of the physical appearance of all those comets which have no tails, or distinct traces of any considerable elongation in the direction of the sun.

The orbit of Tuttle's comet is an ellipse, corresponding to a periodic time of 5005 days, or a little more than thirteen years and eight months. When in the perihelion, its distance from the sun is 100,000,000 miles. Its aphelion distance is 987,-000,000 miles. It passed its perihelion on the 28th of February, 1858, and will return again to the same point about the end of November, 1871.

Another comet of short period is that of *Winnecke*, which was first observed by Pons, at Marseilles, on the 12th of June, 1819. It was observed

22

during a period of only forty days, and from the observations thus obtained, Encke found that its orbit was an ellipse, with a period of only five and a half years. It was so uncertain, however, that no attempt was made to predict its future returns to its perihelion; and, consequently, it was not again seen until the 8th of March, 1858, when it was discovered by Winnecke, at Bonn, and identified by a comparison of its elements with those of the comet of 1819. It was observed until the beginning of June, 1858, and from the series of observations it was found to have a periodic time of 1830 days, or almost exactly five years. It has, therefore, next to Encke's comet, the shortest period which has yet been discovered as belonging to such bodies. Its distance from the sun, when in its perihelion, is about 73,000,000 miles. Its aphelion distance is about 480,000,000 miles. It must, therefore, approach very near the orbit of Jupiter, and in course of time will be subject to excessive perturbations.

This comet is a telescopic object, not unlike Tuttle's comet in its general appearance. It is probably much larger and brighter, but in form bears a close resemblance to it. It passed its perihelion on the 2d of May, 1858, and may be expected to return again in April, 1863.

We have now described those periodic comets whose orbits are supposed to be known with great

precision; but there are others, whose periodic
character is fully as decidedly indicated, but which
have not been observed to a sufficient extent for a
complete investigation of their elements. A revision of the recorded observations of former comets
by modern astronomers, with the aid of the refined
operations of modern mathematical analysis, has led
to the discovery of the great probability of several
among them having revolved in elliptic orbits, with
periods not differing to any considerable extent
from those of the periodic comets already mentioned.
The fact that these comets have not been re-observed
at their successive returns to their perihelia, may be
explained, either by the difficulty of observing them,
owing to their unfavorable positions, and the circumstance of observers not expecting their reappearance, their periodic character not being then
suspected; or because their orbits may have been so
completely changed, by the disturbing action of the
planets, as to keep them continually beyond the
limits of visibility from the earth.

A singular example of this kind is found in the
case of the comet which appeared in 1770, and
which is known as *Lexell's comet,* from the discoverer
of its periodic character. It is, perhaps, the only
instance in which a particular comet is certainly
known to have appeared in the system, making two
revolutions around the sun in an elliptic orbit of

short period, and then disappearing, without furnishing any means of identifying it at any future or
previous return. This comet was discovered by
Messier, at Paris, on the 14th of June, 1770, and
observed till the 2d of October following. It passed
its perihelion on the 14th of August, and on the 1st
of July was distant from the earth only 363 terrestrial radii, or about six times the distance of the
moon.

On computing the elements of the orbit of this
comet, it was found to be impossible to represent
the observations by a parabola. Six years after its
disappearance, Lexell showed that the observations
could be satisfactorily represented by an ellipse,
corresponding to a period of about five and a half
years. This was certainly a remarkable result,
since there was no other example of a comet of less
period than about seventy-five years; and, moreover, the fact that the comet had not been seen at
previous returns, since it was a brilliant object in
1770, seemed to militate strongly against Lexell's
conclusions in regard to its periodic character. The
aphelion distance of the comet was found to be
about 520,000,000 miles, and its perihelion distance,
65,000,000 miles. Its orbit is inclined to the orbit
of the earth by an angle of only a degree and a
half; and it follows, therefore, that it almost literally intersects the orbits of the earth, Mars, and

Jupiter. This fact enabled Lexell to remove all
doubts in regard to the accuracy of his results, since
it could be easily shown that at its previous aphelion, in 1767, the comet must have passed Jupiter
at a distance fifty-eight times less than the planet's
distance from the sun; and that, consequently, it
must then have sustained an attraction from the
great mass of the planet, more than three times
more energetic than that of the sun, at the same
time. It follows, also, that in this manner its former
orbit was changed into an ellipse of only five and a
half years, in which it actually moved in 1770. It
is highly probable that, prior to 1767, its orbit was
a parabola, and that, although moving in an elliptic
orbit from 1767 to 1770, and having the periodicity
consequent on such motion, it nevertheless moved
then for the first time in its new orbit, and had
never come within the sphere of the sun's sensible
attraction anterior to this epoch.

The return of the comet in 1776 could not be
observed, on account of the relative situations of
the earth and comet at that time — the latter being
wholly enveloped in the sun's rays or in the strong
twilight; and before another revolution could be
accomplished, its orbit was again completely
changed. Lexell had further determined that,
since the comet passed through its aphelion, which,
as already stated, nearly intersected the orbit of

22 * R

Jupiter, at intervals of a little more than five and a half years, it having been in that position in 1767, and the period of the planet being less than twelve years, the planet, after a single revolution, and the comet after two revolutions, must necessarily again encounter each other in 1779. He found that the comet would approach Jupiter, about the 23d of August, 1779, within a distance ninety-one times less than the distance of the latter from the sun. It would, therefore, be attracted towards the planet by a force which was at least 200 times greater than the sun's attraction at the same point of its orbit; and, consequently, its orbit would, in all probability, be again changed into a parabola or an hyperbola. It would thus depart for ever, for aught we know, from our system, to visit other suns and systems. Lexell, therefore, announced that in 1770 it was actually describing an elliptic orbit with a period of five and a half years; but that, owing to the excessive perturbations produced by Jupiter in 1779, it might be expected to have already made its final disappearance.

At the time when these computations were made by Lexell, the exact methods of Lagrange and Laplace, for the computation of perturbations, had not yet been developed. Subsequently, the question here propounded was resumed by the latter geom-

eter, and a general solution was obtained for the
following definite problem:

*The actual orbit of a comet being given, what was its
orbit before, and what will be its orbit after being submitted to any given disturbing action of a planet near
which it passes?*

Having obtained a general solution of this problem, he applied his formula to the particular case
of Lexell's comet, and showed that, before sustain-
ing the disturbing action of Jupiter in 1767, the
comet must have moved in an ellipse whose greater
axis was about 2,526,000,000 miles; and, consequently, that its period, instead of being five and a
half years, must have been forty-eight and a half
years. He showed also that the eccentricity of the
orbit of the comet was such that its perihelion distance would be nearly equal to the mean distance
of Jupiter, so that it could never have been visible.
We perceive, therefore, that previous to 1767, the
comet experienced excessive perturbations only when
near its perihelion. It was further demonstrated by
Laplace, that after being subjected to the disturbing
action of Jupiter in 1779, the orbit of five and a half
years was changed into an ellipse whose greater axis
was about 1,400,000,000 miles, thus indicating a period of twenty years. The eccentricity of the orbit
of the comet was found to be such that its perihelion
distance was more than double that of the planet

Mars, so that even in such an orbit the comet could not become visible.

The question rested thus for nearly half a century, when the investigations of Laplace were thoroughly revised by Le Verrier. He found that the observations of 1770 were not sufficiently accurate and numerous to warrant such absolute conclusions; and showed that the orbit of 1770 is subject to an uncertainty comprised between certain definite limits; and also that, tracing the consequences of this to the positions of the comet in 1767 and 1779, these places are subject to still wider limits of uncertainty. Thus he finds that the observations of 1770 can be almost equally well represented, no matter whether the comet is supposed, in 1779, to have passed considerably outside, or considerably inside, of Jupiter's orbit; or whether, as it was supposed to have done, it actually passes within and among the orbits of the satellites. Finally, he deduces the following general conclusions:

1. That if the comet had passed within the orbits of the satellites of Jupiter, it must have fallen down upon the planet and coalesced with it—an incident which must be considered highly improbable, though not absolutely impossible.

2. The action of Jupiter may have thrown the comet into a parabolic or hyperbolic orbit, in which case it must have taken its final departure from our

system, never again to return; except it may be by
the consequence of some disturbance of its motion
produced in another sphere of attraction.

3. It may have been thrown into an elliptic orbit,
having a great axis and long period, and so placed
and formed that the comet would never become
visible—a supposition which agrees with the solution
of Laplace.

4. It may have had merely its elliptic elements
more or less modified by the action of the planet,
without losing its character of short periodicity — a
result which is perhaps the most probable, and which
would render it possible that this comet may still
be identified with some one of the many comets of
short period, which the activity and sagacity of observers are continually bringing to notice.

In order to facilitate all such comparisons, Le
Verrier has given a table, which includes all the possible systems of elliptic elements of short period
which the comet could have assumed, subject to the
disturbing action of Jupiter in 1779, and taking the
observations of 1770 within the limits of their probable errors, but which it will be unnecessary to add
in this connection. He then proceeds to demonstrate
that the orbit in which the comet moved prior to the
disturbing action of Jupiter upon it in 1767, not
only could not have been an ellipse or hyperbola,
but must have been an ellipse whose periodic

time was considerably less than had been deduced
by Laplace, from the insufficient observations of
Messier. His calculations show, that before that
epoch, the perihelion distance of the comet could
not, under any possible supposition, have exceeded
three times the earth's mean distance from the sun,
and was, most probably, included between one and
a half and two times that distance; and that the aphelion distance could not have exceeded six times
the same distance, a magnitude three times less than
that assigned to it by the calculations of Laplace.

Having thus thoroughly investigated the path of
Lexell's comet, both before and after its successive
changes, from the action of Jupiter, Le Verrier went
still further, and endeavored to find whether any of
the periodic comets now known can be supposed to
be identical with this comet. In accomplishing this
it became necessary to trace the paths of each comet
back through all its previous revolutions since 1779,
making due allowance for the disturbances to which
it was subjected by the action of the planets in the
interval which had elapsed, and then to compare
the elements thus obtained for 1779, with the table
of possible orbits of Lexell's comet which had
already been determined. Should one of these
orbits be found to be identical with that of the comet
in question, its identity with Lexell's comet would
be inferred with the highest degree of probability;

but if, on the other hand, such discrepancies were found to prevail as must exceed all possible errors of observation or calculation, the identity could not be assumed. It was found in this way that none of the periodic comets hitherto observed, and more especially those of short period, can be identical with Lexell's comet, however strongly some of the elements of their present orbits might, at first, seem to indicate such a presumption.

Another comet of short period was discovered by Blainplan, at Marseilles, on the 28th of November, 1819, and was observed until the 25th of January, 1820. From the observations which were made during this period of two months, Encke found that its orbit was an ellipse, corresponding to a period of revolution of about five years. Clausen considers this comet to be identical with one which appeared in January, 1743. It passed its perihelion on the 20th of November, 1819, but has not been observed at any succeeding return.

A comet which was discovered by Pigott, at York, in 1783, was shown by Burckhardt to move in an elliptic orbit, with a periodic time of five and a half years. It passed its perihelion on the 20th of November of that year, but has not been seen since the 21st of December following.

On the 26th of June, a telescopic comet was discovered by Peters, at Naples, and observed till the

21st of July. Its orbit has been calculated by D'Arrest, and found to be an ellipse with a period of 5804 days, or nearly sixteen years. Later computations by Peters have assigned to it a periodic time of a little less than thirteen years. It passed its perihelion on the 1st of June, 1846, and was predicted to return again in the first part of 1859. It was, however, in a position very unfavorable for observation in the northern hemisphere, and no observation of it since 1846 has yet been announced.

These are the only additional comets whose period is less than sixteen years, which have hitherto been detected; but there are several with periods of seventy years and upward, whose orbits are supposed to be known with considerable accuracy, although the comets have only been observed at one return to the perihelion. Among these might be noticed a comet discovered by Pons, at Marseilles, on the 20th of July, 1812, for which Encke found a periodic time of about seventy-five and a half years; also, one discovered by Olbers, at Bremen, on the 6th of March, 1815, whose orbit, calculated by Bessel, proved to be an ellipse with a period of seventy-four years. The next perihelion passage of this comet is predicted to take place on the 9th of February, 1887, its time of revolution between 1815 and 1887 being diminished about two years by the action of the planets. A comet, discovered by De Vico, at Rome, on the 28th of February, 1846, is found to have a period of between seventy-two and seventy-three years; and one discovered at Altona, in Denmark, by Brorsen, on the 20th of July, 1847, has been supposed to move in an elliptic orbit with a period of seventy-five years. Still another comet, belonging to the same class, was discovered by Westphal, at Göttingen, on the 27th of June, 1852, and subsequently by Peters, at Constantinople. Its orbit is found to be an ellipse, corresponding to a period of about seventy years.

On the 25th of July, 1857, a faint telescopic comet was discocered by Peters, at Albany, N. Y., which was observed until the 20th of October, following. Soon after its discovery, it was found that its orbit was an ellipse, with a period of about 250 years. Its aphelion distance is about 7,700,000,000 miles, or nearly 5,000,000,000 miles beyond the orbit of Neptune.

These are the only comets of long period whose orbits are known to be contained within the present limits of the planetary system. There are, however, many on record, whose periodicity, although not yet in any instance established by observations made at successive returns to their perihelia, is, nevertheless, so unequivocally indicated, that it is possible to obtain a very approximate value of the dimensions of their orbits, from the observations made during the time of their visibility at a single perihelion passage.

23

We have already given several examples of this class
of cometary bodies in the preceding chapter.

Of all the comets whose orbits have been computed, only twenty-seven have been definitely ascertained to move in elliptic orbits, and thirteen have
been found to move in hyperbolas.. The others
must, therefore, be supposed to have moved in orbits
which were either parabolic, or in which the eccentricity differed so little from that of a parabola as
not to be indicated by the observations which have
been made. Those which move in elliptic orbits
have periods of revolution varying from 3500 to
6000 years, and of these only the following have
been identified and observed at more than one return
to their perihelia.

Name of Comet.	Period of revolution in days.	Time of next return to the perihelion.	
Halley's	27877	February,	1912
Encke's	1207	"	1862
Biela's	2421	"	1866
Faye's	2727	March,	1866
Brorsen's	2031	October,	1862
D'Arrest's	2340	July,	1864
Tuttle's	5004	November,	1871
Winnecke's	1830	April,	1863

A very curious circumstance, which has often been
noticed in connection with the periodic comets, is
that those already known may be arranged, for the
most part, into two classes — one including those
comets whose mean distances from the sun are all

nearly the same with those of the small planets between Mars and Jupiter; and the other class, including those whose mean distance is, in every case,
very nearly equal to that of Uranus. It has, therefore, been conjectured that the comets composing
the first class may have formerly constituted one, or
at most two, whose orbit or orbits were contracted
by the disturbing action of Jupiter, and that afterward they may have been sub-divided—forming the
different comets which belong to this class — by a
process yet unknown, but probably similar to that
by which Biela's comet has been separated into two
distinct parts almost under the eyes of the observers.
This is the theory advanced by Professor Stephen
Alexander, who imagines that the catastrophe or succession of catastrophes which divided these comets,
must have been of a very ancient date. The comets
of Peters (1846) and Tuttle, whose periods are a
little more than thirteen years, may perhaps be an
exception to any general law or order of distance
which may thus be supposed to exist.

The existence of a group of small planets between
Mars and Jupiter has long since been accounted for
by the supposition that they at one time, although,
perhaps, almost infinitely remote, constituted a single
planet which revolved around the sun at about the
average of their mean distances. This original
planet is supposed to have been broken into almost

innumerable fragments by some internal convulsion
of nature, thus forming a group of small planets or
fragments revolving at different distances, but within
the orbits of Mars and Jupiter. In this way then
the common origin of the asteroid planets may be
considered as conclusively established. It has, accordingly, been conjectured that since the comets
of short period may be supposed to have a common
origin also, the latter may have formed from portions of the original planet; and that thus both the
asteroids and comets of short period may have
formerly constituted but one mass, which was subsequently divided by some unusual and violent physical convulsion. When the cohesion of the particles
of matter which constituted the original mass, was
overcome by the action of the explosive force, the
fragments which would thus be thrown off, would
be projected with a velocity depending on their
magnitude and density. The largest fragments
would continue to move on in orbits differing but
little from that of the original planet, while the
lighter fragments would assume orbits more or less
eccentric, depending on the velocity of projection.
If now we suppose that the planet was, just before
its dissolution, composed in part of matter in a
gaseous form, we may conclude with some degree
of probability, perhaps, that the comets of short
period may have been formed in this manner. The

COMETS AND ASTEROID PLANETS. 269

very difference of inertia of the different parts of the planet, might seem to have caused the separation, under these circumstances, of the nebulous or cometary matter, from that of greater density; and while portions of this same nebulous material were still retained by the larger portions of the original mass, the smaller portions would retain scarcely any traces of it. In support of this hypothesis, the fact has been cited that some of the asteroid planets are actually observed to be surrounded by a very dense atmosphere, or nebulous envelope.

However plausible these hypotheses in reference to the similarity and common origin of the asteroid planets and the comets of short period may seem to be, yet we cannot by any means regard them as having anything more than the very slightest degree of probability. There is a very beautiful theory of the formation of the solar system,—which we shall explain hereafter,—by which the origin of cometary worlds is illustrated with great distinctness, and which presents far greater claims to our attentive consideration than any other which has ever been devised. Reference is here made to the nebular hypothesis of Laplace, who, in his theory of planetary genesis from rings of vapor revolving round the sun, in which matter aggregates into spheres around a nucleus, asserts that, according to the hypothesis of zones of vapor, and of a nucleus increasing by

23 *

the condensation of the atmosphere which surrounds them, the comets are to be regarded as strangers to the planetary system.

When we consider the subject of the comets in its fullest extent, it may indeed be said that there is no other branch of astronomy more replete with interest, as furnishing the most sublime examples of the achievements of the human intellect in the investigation of the intricate problems here presented. It is certainly wonderful to notice the extreme precision of the calculations of modern astronomers regarding the orbits of these chaotic worlds, by which they are enabled to conclude definitely what has been the past, and what will be the future history of any particular body. Each newly discovered comet is at once subjected to the ordeal of a most rigorous inquiry; and its elements, roughly calculated within a few days after its first appearance, are gradually corrected as observations accumulate, until finally the exact form and position of its orbit in space are determined. Its connection or identity with any comet which may have previously appeared is carefully examined, and when a decided similarity has been detected, the disturbances due to the action of the planets during the interval which has elapsed, are investigated; and the past, thus brought into unbroken connection with the present, is made to afford substantial ground for prediction of the future.

The increasing attention which is being bestowed upon this branch of knowledge, may ultimately increase the number of periodic comets to an extent almost beyond conjecture; and when, even now, we come to consider the manner in which the orbits very nearly intersect those of the various planets belonging to our system, it may indeed seem probable that the time will come when some one of these bodies will come into collision with some member of the planetary system. How far the welfare of the inhabitants of the earth may be dependent upon any such event, when the collision is not directly with our planet, it will be unnecessary to attempt to determine; and let it suffice to know that calculations, based on the theory of probabilities, and of the most positive character, show that there is but one chance in 281,000,000, that the earth will come in collision with one of these bodies. The probable effect of such a catastrophe, admitting its possibility, has already been discussed in a previous connection. For some of the planets, the probability of coming in contact is much less than in the case of the earth; and for others it is even greater. It has indeed been asserted that Mars was struck by a comet in 1315 or 1316, or that there was, at least, an appulse, if not an actual collision. What may have been the effect produced on the comet, in case this event did actually take place, we have no means

of determining; but in respect to Mars, there is no
indication in its motions, or in those of the neigh-
boring planets, that it has ever suffered disturbance
from any such source.

It might also be expected, considering the great
number of comets which are supposed to be included
within the solar system, that a collision would occasionally take place between two of these bodies.
The orbits of Encke's and Biela's comets actually
intersect each other, and it is highly probable that
at some future time they may meet in this point.
Should this event take place in the month of October, it is possible that the inhabitants of the earth
may witness the extraordinary spectacle of an encounter between these two bodies; and, it may be,
of their reciprocal penetration and amalgamation,
or of their destruction by means of exhausting emanations. In view of this fact, it has most justly
been stated that events of this nature, resulting
either from deflection occasioned by disturbing
masses or primevally intersecting orbits, must have
been of frequent occurrence, in the course of millions
of years, in the immeasurable regions of ethereal
space. Still, they must be regarded as isolated oc-
currences, exercising no more general or alterative
effects on the order and harmony of the celestial
motions, than the breaking forth or extinction of a
volcano, within the limited sphere of our earth.

CHAPTER IV.

We have already stated briefly the general characteristics by which the comets are distinguished from the other members of the solar system, and have also explained the terms by which each part of a comet is designated. It is proposed, therefore, in this connection, to consider more fully the physical constitution of these wandering bodies, the effect of the various degrees of temperature to which such objects may be subjected, and, finally, their probable origin. The theory of their motions we have found to be in every respect complete, but in the case of their physical constitution it is far different. The data here presented are by no means of a positive character, and the conclusions which may be derived must be received with due allowance. There may, indeed, be comets in existence which bear some slight resemblance to the planets in their physical constitution, and which, were it not for the great eccentricity of their orbits, might justly be regarded as planetary worlds. But by far the greater number of the comets, and more especially those which are visible only through a telescope, are seen generally as clouds or masses of vapor, with or without a train of fainter light.

It has long since been noticed that comets might pass directly between the eye of the observer and stars of the sixth and seventh magnitude, without diminishing their light; and that, in every such case, the effect produced on the light of the star would not be greater than might be expected if it were shining through a very slight fog. There are numerous cases of this kind on record, all of which tend to show that the comets are composed of matter in a state of the very greatest attenuation. From considerations such as this, and the fact that the light of the stars are greatly enfeebled by that of the full moon shining through our atmosphere, attempts have been made to determine the probable masses of some of the comets; and the results obtained have invariably indicated that they contain but a very small quantity of matter. In some cases it has been found that the absolute quantity of matter in the entire comet was almost inconceivably small; and that, consequently, the comet was so extremely attenuated that the molecules of which it is composed were completely isolated, and destitute of mutual elastic reaction. These results, however, must be received with considerable allowance for the insufficiency of the data on which they are based. It is by no means certain that a central passage of a comet over a star will not almost invariably cause a complete occultation. In those cases in which stars have been seen through comets, they have usually been situated barely within the exterior nebulosity or envelope, and, consequently, not under circumstances favorable to indicate the precise density of the cometary matter. In April, 1857, the complete occultation of a star by Brorsen's comet was observed at Florence; and a few months later the same phenomenon, produced by another comet, was observed at Altona, in Denmark. In both cases the comet was a telescopic object, and belonged to the class of diffused nebulosities. Thus, although there were no indications that the comet was really solid at the centre, yet the density of the nucleus was such as to prevent a passage of the light of the star. There can be no doubt but that many of the comets have actually a solid nucleus surrounded by an immense nebulous atmosphere. The solid part, however, may not exceed a few miles, or perhaps a few hundred miles, in diameter, while the nebulous envelope, or cometic atmosphere, may be of such enormous dimensions as to appear as if constituting the whole matter of the comet. In this case the comet might pass almost centrally over a star and yet not affect its brilliancy very materially, or cause a complete occultation. It is evident, therefore, that the combination of circumstances which would alone furnish a sufficient test of the density of a comet, by passing apparently over a star, is of rare occurrence. It would be necessary that the centre of the head of the comet, although very small, should pass critically over a star, in order to ascertain whether the latter is visible through it.

In September, 1832, Biela's comet was observed to pass directly over a small cluster of telescopic stars of the sixteenth or seventeenth magnitude; and the entire cluster remained distinctly visible, thus affording a striking proof of the extreme translucency of the matter of which this comet consists. The most trifling fog or haze in the earth's atmosphere would have entirely effaced this group of stars, yet they continued visible through a thickness of cometic matter which, at its central part, must have been nearly 50,000 miles. In this case, therefore, the comet could not have had a solid nucleus of any considerable size, but it does not follow that a similar phenomenon could have been observed as a general result for all comets. It is certainly known that many of the comets which have hitherto appeared have a solid and compact nucleus; and we may, therefore, conclude, that while there are many comets without any nucleus, properly so called, there are some with a nucleus which, perhaps, may be transparent ; and others more brilliant than planets, having a nucleus which is solid and opaque. In those comets in which the appearance of a nucleus is most distinctly marked, the nebulous envelope, instead of a uniform or of a progressively increasing brilliancy from the circumference toward the centre, is often apparently composed of one or more rings or strata, many thousand miles in thickness, alternating with fainter portions—thus presenting an appearance not unlike what would be presented to a spectator viewing from a distance our earth with its different layers of clouds, one above another, separated by portions of transparent atmosphere.

It appears from such considerations that, as a general result, the absolute quantity of matter contained in any of the comets which have hitherto appeared, must be extremely small when compared with that of the planets belonging to our system, excepting, perhaps, some of the group of asteroid planets between Mars and Jupiter. That they are thus constituted is still further evident from the fact, that for ages they have travelled among the planets and their satellites without producing any disturbances which have yet been detected. An example of this kind has already been given in the case of

24

Lexell's comet of 1770, which passed directly among the satellites of Jupiter, without in the least disarranging that system. The comet, however, was subjected to the most excessive perturbations by Jupiter, its orbit being completely changed. We have also numerous instances on record, in which comets have approached very near some of the smaller planets of the system without producing the slightest perceptible disturbance, but in which, without exception, the comet has been drawn very considerably from its previous orbit. According to the law of universal gravitation first discovered by Newton, and subsequently demonstrated to the very fullest extent, as the great law or bond which unites the various bodies composing the material universe, by the researches of Lagrange, Laplace, and others, every portion of matter attracts and is attracted directly as the quantity of matter, and inversely as the square of the distance from the attracting body. This is the law of universal gravitation; and it may be perceived at once, from the mere statement of the law, that the comets must be greatly attracted from their regular course by the various bodies of the systems through which or near which they pass. The disturbances to which they are thus subjected, and to which we have briefly alluded in the preceding chapter, are called the *perturbations*.

To attempt an explanation of the manner in which

the perturbations of the comets may be determined, would require a higher degree of mathematical knowledge than can be expected of those for whom these explanations are intended; and we must therefore be contented with a mere statement of the general features of the problem thus presented. If the system were composed of simply the sun and one planet, the latter would describe an exact ellipse around the former, or rather both around their common centre of gravity, and continue to perform its revolutions in an unchanged orbit through all time. But if now we add to the system a third body, the attraction of this third body will draw both of the former bodies out of their mutual orbits, and by acting on them unequally, will disturb their relation to each other, and will completely efface all traces of exact elliptic motion either around one another, or about a fixed point in space. The disturbance thus produced is not due to the whole attraction of the third body, but rather to the difference of its attractions on the two which we have supposed to have originally constituted the system.

In the solar system, when compared with the sun, the masses or weights of the different comets and planets composing the system, are extremely minute. Their attractions on each other, therefore, are all very feeble when brought into comparison with the presiding central power of the sun, and the effects of their disturbing forces are proportionally small. This fact enables us, in determining the perturbations of any particular body, to estimate each of the effects of the disturbing bodies separately, as if the others did not take place, without causing any error in the final results beyond what may be considered necessarily incident to a first approximation. It follows directly from the primary relations between forces and the motions which they produce, that when a number of very minute forces act at once on a system, the combined effect is the sum or aggregate of their separate effects, at least within such limits, that the original relation of the parts of the system shall not have been materially changed by their joint action. This, in the case of the mutual attractions of the planets and comets, may be supposed to be true for a limited portion of time, but when the orbits may have become slightly changed, it becomes necessary to take account of these changes in determining the subsequent effects. It is evident, however, that in investigating the disturbing influence of the several planets on any particular comet, we need only consider the case of a single disturbing planet at a time, and having found their individual effects, the sum of these will be the total effect. The problem thus reduces to that of three bodies; a predominant central body, corresponding to the sun, a disturbing body, corresponding to a planet, and a disturbed body, or the comet. It is here supposed that the mass of the comet is small when compared with either that of the sun or planet, which has hitherto invariably been the case; and it may, perhaps, be stated as a general fact, that no comet will probably ever appear which will render it necessary to take into account its action on any of the planets.

Since the orbits of the comets are inclined to the orbit of the earth at all possible angles, while the orbits of the other planets are all inclined to this within very narrow limits, excepting, of course, some of the asteroid planets, it follows that for different comets the effect produced by the several planets on the different elements will be greatly dissimilar. For example, if a comet happened to pass near a planet when in the vicinity of the node of the former, the attraction of the latter might produce a great change in the inclination of its orbit. If the comet, on the contrary, were far distant from either node of its orbit, the effect would be to change the place of the node without affecting very considerably the inclination. In the case of a comet whose orbit is greatly inclined to the orbit of the earth, since its distance from the planets in the remoter parts of its orbit will be much greater than if it were less inclined, the perturbations will be proportionally smaller. It is evident, therefore, that those comets

24 *

whose orbits have the greatest inclination to those
of the planets, are subject to the least disturbance;
while those which move very nearly in the plane of
these orbits, or else in that which would result as a
mean between the orbits of the principal planets,
would, on account of the great length of their orbits—
which would bring them into the vicinity of several
different planets — be subject to much greater disturbance.

In computing the perturbations of a comet by a
planet, the attractive force of the planet on the
comet, depending, of course, on their mutual distance, is resolved into three parts called *components*,
each acting in the direction of one of the co-ordinate axes, already explained in the preceding chapter.* The amount of the disturbance of the motion
of the comet in the direction of the three rectangular
axes is thus obtained; and it is only necessary to
add them to the rectangular co-ordinates of the
comet computed under the supposition of an undis-
turbed orbit, or subtract them from these, as the
case may be, in order to find the actual co-ordinates
of the comet at any given instant. In this way the
perturbations are to be computed from day to day,
or, when they are not very large, at intervals of
twenty or forty days, or even longer, correcting the
elements of the orbit, from time to time, by means

* See page 170.

of the variations of the co-ordinates already obtained. Assuming then the elements of the orbit to be accurately determined from observations for any given epoch, the perturbations or variations of the co-ordinates must be computed at regular intervals, as above stated, for the whole period during which it is required to trace the motions of the comet. The same process must be performed for each disturbing planet, and the sum of the separate variations for each planet will give the total amount of the perturbations. Having obtained these values, and having corrected the assumed elements accordingly, we obtain finally the elements of the orbit in which the comet is moving at the instant for which its true orbit is required. The values of the disturbances thus formed are called the *special* perturbations of the comet. In the case of the planets it is possible to compute the variations of their elements due to their mutual attractions for an indefinite number of revolutions. The formulæ, however, which give these variations, are so affected by the eccentricity of the orbit of the planet, that the solution is possible only when the eccentricity is small. The variations of the elements thus obtained are called the *general* perturbations, and can be computed only in the case of the planets — the eccentricity of their orbits being small. In the case of the comets the eccentricity of the orbit is always so great that it is

impossible to compute their general perturbations,
or for an indefinite number of revolutions; and,
consequently, astronomers are enabled to determine
only the variations of the elements of their orbits—
assuming the elements for a given epoch to be
known—for a series of dates in regular progression.
When a comet passes to such an enormous distance
that the attraction of even the remote planets of
the system becomes insensible, it is no longer necessary to compute the variations of its elements as
before, since it may then be supposed to revolve in
a pure ellipse around the common centre of gravity
of the system, which, however, coincides very nearly
with the centre of the sun. It is supposed to revolve
thus in an orbit which may be determined from the
assumed elements in connection with their variations
or perturbations as already determined, until finally
it again comes within the sphere of the sensible
attraction of the planets, when the disturbance of
its motion is again computed for a regular series of
dates, precisely as before. In this way, therefore,
the computation of the perturbations of a comet
may be performed for a long series of years, or even
for a number of revolutions; but the process, as may
be readily perceived, is very laborious. In the case
of the periodic comets the perturbations are computed only to the date of their next appearance,
since it would be useless to continue them further

without correcting the elements by means of a series
of observations which might then be taken. In the
case of a comet moving in an ellipse of very great
eccentricity, or in which the period is more than
1000 years, and also in the case of a comet moving
in a parabola or hyperbola, it is necessary only to
compute the perturbations during the short period
in which it is visible from the earth, since in these
cases all that is required is to find the exact elements
of the orbit in which the comet was moving while
visible.

Such is a general statement of the method by
which astronomers are enabled to follow the comets
in their long journeys through space, while invisible
to the inhabitants of our earth, and to predict, in
the case of the periodic comets, the exact period of
their return. To be able to solve this problem
requires the very highest degree of mathematical
knowledge; and let it suffice here to say that it is
one of the most difficult problems of astronomy,
while the amount of labor required in making the
numerical computations, vastly exceeds what may
be readily conceived. An illustration of this fact
may be witnessed in the case of Halley's comet
already given.

We have already stated that in all computations
respecting the motions of the comets, their masses
have been neglected as being inconsiderable in com-

parison with those of the sun and planets. Their
effect in the planetary system, even if there were
not other facts tending to the same result, would
alone be sufficient to indicate that this supposition is
entirely admissible. These bodies are continually
passing through our system, and often in the immediate vicinity of or directly among the asteroid
planets, and yet do not cause the slightest disturbance which can be determined by accurate observation. The physical appearance of the comet affords
also conclusive proof that their masses are at least
small. Various attempts have been made to determine the masses of these bodies; and although the
results obtained exhibit great discrepancies, yet they
all indicate conclusively that in the case of a very
great majority of the comets the mass is extremely
small, or, in reality, almost inconceivably small.
It may, perhaps, seem really absurd to assert that
the tails of some of the largest comets have not
contained a hundred pounds of matter, yet it has
been asserted that even the largest and longest did
not contain more than a few ounces. We are, of
course, unable to deny such an assertion, since we
have no means of proving the contrary. It must be
admitted, however, that this is an extremely small
estimate, and that, although the comets are in reality
of extreme tenuity, and consequently have no considerable mass, yet that they contain vastly more

matter than what would result from any such estimate. The only correct and reliable method of
arriving at the exact value of the mass of a comet,
is by means of the disturbance which it may produce
in the motions of a planet near which it approaches;
and until such a phenomenon is observed, no definite
knowledge of this element can be obtained.

When we come to consider the ever-varying temperature to which the comets are subjected in their
motions through space, the question may very
naturally arise as to the probable effect which would
thus be produced. It might be supposed that, as a
necessary consequence, when a comet approaches
the sun, the continually increasing temperature
which may be supposed to affect it, would cause it
to expand gradually until it arrived at its perihelion,
when its absolute dimensions would be greatest.
For a similar reason it might be expected that in
receding from the sun, the temperature being supposed to decrease continually, it would be gradually
contracted, and would finally appear precisely as
before its perihelion passage. It may, on this account, seem strange that the dimensions of a comet
are observed almost invariably to be enlarged as it
recedes from the source of heat, and more especially
in the case of those telescopic comets which have no
tail. This singular and somewhat unaccountable
phenomenon has been explained in various ways.

Valz ascribed it to the pressure of the solar atmosphere, supposed to extend to a great distance, acting upon the comet. He supposes that the atmosphere, being more dense than the sun, compressed the comet and diminished its dimensions; while, at a greater distance, being relieved from this coercion, the body again expanded to its natural bulk. In order to test the validity of this hypothesis, probable values were assumed for the density of the solar atmosphere and the elasticity of the comet; and the variations of the bulk of a comet, deduced in accordance with this assumption, exhibited a remarkably close agreement with the observed change in its dimensions. It is necessary, however, to suppose also that the comet is composed of an elastic gas or vapor; and, further, that it is impervious to the solar atmosphere through which it moves, both of which assumptions are inadmissible, at least in the generality of cases. Another theory which has been advanced in order to explain the fact that the dimensions of a comet are enlarged as it recedes from the source of heat, is, that as the particles or molecules of the nebulous matter of the comet are distant, and held together by so feeble a power, they may revolve to a certain limited extent independently of each other, each having its own perihelion; and that thus they would be brought nearer to each other as they approach the sun, and separate again further and further as they depart from him.   This theory assumes that the mutual cohesion, or mutual gravitation of these particles is a quantity evanescent in comparison with their separate gravitation toward the sun.

Sir John Herschel supposes that the nebulous portion of the comet, or that portion which reflects the sun's rays, is of the nature of a fog, or of a collection of discrete particles of a volatile fluid floating in a transparent medium.   During the comet's approach to the sun, these molecules would absorb its rays and become heated, and consequently a portion of them would be constantly passing from the liquid to the extremely gaseous or invisible state, just as a fog or common vapor disappears before the rising sun.   As this change must commence from without, and must be propagated toward the centre, the ultimate effect would be a diminution of the visible bulk of the comet.   As the comet receded from the sun, it would lose by radiation the heat thus acquired; which, in accordance with the general analogy of radiant heat, might be expected to escape chiefly from the unevaporated or nebulous mass within. The dimensions of the latter would, therefore, begin and continue to increase by the precipitation above it of fresh nebula, just as fogs, on cold still nights, are seen to form at first on the surface of the earth, and as the heat near the surface becomes dissipated.

25   T

gradually extending upward. A comet might thus appear to enlarge rapidly in its visible dimensions, while its real volume was slowly contracting by the general abstraction of its heat. Herschel supposes that this process might go on in the absence of any solid or fluid nucleus; and in those cases in which such a nucleus exists, the increase of temperature in the vicinity of the sun, by causing an evaporation from its surface, would afford a constant and copious supply of vapor, which, rising into its atmosphere, and condensing at its exterior parts, would tend still more to dilate the visible limits of its nebulosity. In this manner we may account for the appearances which have been noticed in the head of certain comets, where a stratum void of nebula has been observed, interposed, as it were, between the denser portion of the nucleus, and the coma or envelope. Another theory which has been advanced by Herschel, in explanation of the phenomenon under consideration, attributes it to the ethereal medium surrounding the sun. It is supposed to be by no means improbable that the region in which the earth revolves has a temperature of its own greatly superior to what may be presumed to be the absolute zero, and even to what may be produced by artificial means. This temperature Herschel supposes to be due not simply to the radiation of the stars, but rather to the contact of an ether, possessing itself a determinate temperature, and tending, like all known fluids, to communicate this temperature to bodies immersed in it. Now, if the temperature of the ether increases in approaching the sun, which seems to be a necessary consequence — regarding it as endued with the ordinary relations of fluids to heat — an obvious explanation is furnished of the phenomenon under consideration. A body of such extreme tenuity as a comet, may be presumed to assume very readily the temperature of the ether by which it is enveloped; and the vicissitude of warmth and cold thus experienced may ultimately convert it into transparent vapor, and again precipitate the nebulous substance, just as fogs are dissipated by an increase of atmospheric temperature, not by abstracting or annihilating its aqueous particles, but by causing them to assume the elastic and transparent state, which, when the temperature falls, they again lose, and appear in the formation of another fog.

Another important consideration immediately connected with this branch of our subject, is whether the comets shine partially by their own proper light, or, in other words, whether they are to be regarded as to a limited extent self-luminous. If a comet should have a solid and opaque nucleus of any considerable size, it would follow as a necessary consequence, in case it did not have a light of its own, that it would exhibit the phases of the moon when near the sun. The fact, however, that these have rarely, if ever, been certainly observed, and also that the comets are known to be composed of nebulous matter in an extremely attenuated condition, has rendered it uncertain whether they shine by reflected light, like the planets, or by virtue of their being self-luminous, like the sun and stars. A mere mass of nebulous matter, not itself luminous, but rendered visible by reflected light, would not exhibit phases similar to those of the moon and interior planets. Its imperfect opacity would permit the solar light to affect its constituent parts throughout its entire depth — so that, like a thin, fleecy cloud, it would not appear as if superficially illuminated, but as receiving and reflecting light through all its dimensions, both externally and internally. If we suppose the comets to be self-luminous, it might be expected that they would be visible in the remote parts of their orbits until their distance was absolutely so great as to render their apparent diameter inappreciable. It is an established property of self-luminous bodies, that when viewed from any distance whatever, they will appear of the same intrinsic splendor, or, in other words, equal portions of the apparent surface at different distances will be equally brilliant. Thus the sun, when seen from Neptune, the exterior planet of our system, must appear as bright as when seen from the earth. The absolute quantity of light is much smaller, since the great increase of distance from the sun in passing from the earth to that planet will cause the apparent angular diameter of the sun to be proportionally diminished. It is evident, therefore, that although equal portions of his disc, as seen from these planets, may be equally bright, yet the total amount of light received at each will be vastly different. It follows from this, that, since the visibility of a self-luminous object does not depend upon the angle which it subtends as long as it is of sensible magnitude, if a comet shines by its own light, it should retain its brilliancy as long as its diameter is of a sensible magnitude; and that even after it has receded to such a distance as to render its apparent diameter inappreciable, it ought to be visible like the fixed stars, and should only vanish in consequence of extreme remoteness. The phenomena, however, which are actually observed, do not in the least accord with any such hypothesis. The comets are observed to become gradually fainter and fainter as their distance from the earth and sun increases, and finally become invisible, even through the most powerful telescopes, while their apparent angular diameter is very considerable. This would seem to indicate conclusively that they shine by reflected light. It might be remarked also, in this

25 *

connection, that the brilliancy of a comet computed from day to day, during the period of its visibility, under the supposition that it shines by reflected light, is found to coincide very closely with the actual appearance of the comet. The most brilliant comets have ceased to be visible long before they had receded to a distance of 500,000,000 miles; and a very great majority of these bodies have their perihelia within the orbit of Mars, while none have been visible from the earth whose perihelia were exterior to the orbit of Jupiter.

The facts thus established are entirely incompatible with the hypothesis that the comets are self-luminous bodies, unless it is supposed that, from some physical cause, they gradually lose their luminosity. The phenomenon, already noticed, of the gradual expansion of a comet in receding from the sun, might perhaps serve to account partially for a diminution of their luminosity. The luminous matter thus existing in a less condensed state might be supposed to shine with a proportionally enfeebled splendor, until, at length, by the dilatation of the body, the light would become so reduced as to be incapable of affecting the retina of the eye to an extent sufficient to produce a sensation. The expansion of the comet, however, would have precisely the same effect in the case of reflected light; and it is evident, therefore, that nothing definite can be

arrived at in this manner, except by experiment.
Arago has submitted to examination the rate in
which comets increase their dimensions in receding
from the sun, and has found the corresponding
diminution of intrinsic splendor which would arise
from such a cause. It remains then to determine
whether the circumstances are such as to render the
brightest comets visible beyond the orbit of Jupiter.
This Arago proposed to determine by actual experiment. Let a telescope be selected having a large
aperture and low magnifying power, by the aid of
which the comet may be observed in every part of
its visible course. Let the comet be observed at
some determinate distance from the sun, and by
varying the magnifying power of the telescope, it
may be made to assume apparently different degrees
of brilliancy. The magnifying power of the telescope may be so regulated as to exhibit the comet
with precisely that degree of brilliancy with which
it would appear at any given increased distance
from the sun, if observed with the lowest magnifying power of the instrument, under the supposition
of its being a self-luminous body, and losing brightness by reason of the enlargement of its dimensions.
In this manner the actual brilliancy of the comet,
if self-luminous, at any given distance from the
sun, might be predicted; and, in case the subsequent observations were found to agree completely

with the prediction, it might be presumed that the comet was visible by means of its own light. But if, as the observations actually indicate, the brilliancy of the comet at different distances from the earth is greatly less than in accordance with the prediction, while it becomes invisible at distances at which its apparent diameter is considerable, then it may be considered as conclusively established, that the body is not self-luminous, but that it derives its light from the sun, and that its final disappearance from our view arises from the extreme faintness of the light reflected by its attenuated matter.

It was the opinion of several astronomers that the great comet of 1811 shone by inherent light, and it was asserted that the rapid variations which were observed to take place in the brilliancy of the nucleus, together with the flashes of light which characterized the appearance of the tail, cannot be explained by means of any other hypothesis. In opposition to this opinion, it may be remarked that comets have been seen as dark spots crossing the sun's disc, and also that their light exhibits traces of polarization. On the 18th of November, 1826, the transit of the dark body of a comet across the sun's disc was observed at Marseilles and Viviers, in France; and a similar phenomenon was observed at various places on the 6th of June, 1818. There are also numerous cases on record in which dark

bodies have been observed in their transit over the
sun's disc, which were undoubtedly comets having
a solid and opaque nucleus. In these cases the
comet could not have been self-luminous, since it
would then have been invisible; or, at least, would
have appeared as a bright instead of a dark spot.
Again, it is found that all direct light constantly
divides itself into two points of the same intensity
when it traverses a crystal possessing the power of
double refraction; while reflected light, in certain
positions of the crystal through which it is made to
pass, gives two images of unequal intensity, in case
the angle of reflection is not ninety degrees; or, in
other words, it is polarized in the act of reflection.
In general, when a ray of light is reflected from any
surface, it may be reflected a second time from
another surface, and may also be made to pass freely
through transparent bodies. If, however, a ray of
light be reflected from any surface at a given angle,
—to be determined by experiment,—it will be found
to have been rendered totally incapable of reflection
at another surface in certain definite positions;
while, in other positions, it will be completely reflected by the second surface. It will be found also
to have lost the property of penetrating transparent
bodies in particular positions, but, in others, to be
freely transmitted by them. Light thus modified
so as to be incapable of reflection and transmission

in certain directions, is said to be polarized. On this principle Arago proposed a photometrical method of determining whether the comets are self-luminous, or shine by means of the light of the sun reflected from their surface, and, perhaps, even their internal parts. This method was applied at the last return of Halley's comet, and it was found that the apparatus gave two images presenting the complementary colors, one of them being red and the other green. By turning the instrument half round the red image became green, and the reverse. It was therefore concluded that the light of the comet, or at least the whole of it, was not composed of rays possessing the property of direct light. The same experiments were repeated with this comet by different observers, and always with the same result. Similar experiments have subsequently been made on some of the brightest comets, and it is found that their light exhibits precisely the same phenomena. This shows that the light of comets must be partly composed of that received from the sun, even if they have a light of their own. All that can be proved by the polarization of their light, is that they may possibly shine wholly by reflection, while the considerations previously adduced are sufficient to establish the possibility thus admitted as a physical fact. In view of all these facts it seems reasonable to conclude, without hesitation, that the comets shine by means of the light which they receive from the sun, reflected from all their parts.

The next important consideration connected with the theory of the physical constitution of comets, is that which relates to the formation and development of their tails. The idea of a tail or train of vast extent is, in the popular mind, inseparably connected with the physical appearance of these wonderful bodies; but it should be remarked, that by far the great majority of comets are not thus attended. Those which have been so brilliant as to be visible to the naked eye, have usually exhibited the phenomenon of a train of fainter light, extending in a direction nearly opposite the sun. The direction of the tail is, however, by no means invariable. Sometimes it has happened that it formed a considerable angle with a line drawn to the sun, and cases have occurred in which it was actually at right angles to it. It has also been observed that the tails of comets, as a general result, incline constantly toward the region from which the comet is moving, as if, in its progress through space, it were subject to the retarding influence of some resisting medium, its nebulous atmosphere being, on account of the resistance thus imposed, compelled to remain behind the solid nucleus in the form of a tail.

We have already, in the descriptions of some of

the most remarkable comets given in the preceding pages, had occasion to allude to the curved form and absolute dimensions of the tails of comets, and have also noticed the various other characteristics by which these somewhat remarkable appendages are distinguished. It is proposed, therefore, in this connection, to give simply some of the various theories which have hitherto been advanced in explanation of this phenomenon. It will, of course, be useless to attempt to give all the theories which human ingenuity has devised; and we shall, therefore, explain only those which have been received with the greatest favor, before proceeding to illustrate that which is now satisfactorily ascertained to be the true theory of the development of these anomalous appendages.

The more ancient theory of the formation of the tails of comets, in which they are supposed to be formed by the lighter parts being thrown off by the resistance of the ether through which the comet passed, has just been alluded to. In more modern times the prevailing opinion has been, that these particles are driven off by the impulsive force of the sun's rays. The fallacy of the former opinion is evident from the fact that the tails are generally directed from the sun; and that, whatever the relative direction may be, it remains very nearly the same both before and after the perihelion passage.

The more modern theory may be supported with considerable force of reasoning; yet the latest investigations have indicated beyond a doubt the true theory of the formation of the tail. Another theory which has been advanced supposes the tail to be formed by the sun's rays slightly refracted by the nucleus in traversing the envelope of the comet, and uniting in an infinite number of points beyond it, throwing a stronger than ordinary light on the ethereal medium, near to or more remote from the comet, as the ray is more or less refracted from its relative position and direction.

The phenomena presented by Halley's comet at its last appearance, when considered in detail, led Sir John Herschel to conclude: 1st. "That the matter of the nucleus of a comet is powerfully excited and dilated into a vaporous state by the action of the sun's rays, escaping in streams and jets at those points of its surface which oppose the least resistance, and in all probability throwing that surface, or the nucleus itself, into irregular motions by its reaction in the act of so escaping, and thus altering its direction. 2d. That this process chiefly takes place in that portion of the nucleus which is turned toward the sun — the vapor escaping chiefly in that direction. 3d. That when so emitted, it is prevented from proceeding in the direction originally impressed upon it, by some force directed *from*

26

the sun, drifting it back, and carrying it out to vast
distances behind the nucleus, forming the tail, or so
much of the tail as can be considered as consisting
of material substance. 4th. That this force, whatever its nature, acts unequally on the materials of
the comet, the greater portion remaining unva-
porized, and a considerable part of the vapor actually produced remaining in its neighborhood, forming the head and coma. 5th. That the force thus
acting on the materials of the tail cannot possibly
be identical with the ordinary gravitation of matter,
being centrifugal or repulsive, as respects the sun,
and of an energy very far exceeding the gravitating
force toward that luminary. This will be evident
if we consider the enormous velocity with which
the matter of the tail is carried backward, in opposition both to the motion which it had as part of the
nucleus, and to that which it acquired in the act of
its emission, both which motions have to be destroyed in the first instance, before any movement
in the contrary direction can be impressed. 6th.
That unless the matter of the tail thus repelled from
the sun be retained by a peculiar and highly energetic attraction to the nucleus, differing from and
exceptional to the ordinary power of gravitation, it
must leave the nucleus altogether; being in effect
carried far beyond the coercive power of so feeble
a gravitating force as would correspond to the

minute mass of the nucleus; and it is therefore very conceivable that a comet may lose, at every approach to the sun, a portion of that peculiar matter, whatever it be, on which the production of its tail depends, the remainder being of course less excitable by the solar action, and more impassive to his rays, and therefore more nearly approximating to the nature of the planetary bodies."

The repulsion which is thus assumed to be exercised by the sun, was exhibited by Donati's comet in the most palpable manner. The question may now arise whether this force is real or apparent; whether, if real, it is a polar force, like magnetism and electricity, or a simple force like that of gravity; and whether, if apparent, it results simply from the difference of action of the solar attraction on the various parts of the comet, or from its different action on the molecules of a gravitating ether, and upon those of the comet. There are thus exhibited four cases to be considered, two of which, however, may hardly with propriety be regarded as strictly connected with a repulsive force, such as will be proved to exist. The idea of an apparent repulsion is to assimilate the phenomena of the comets to that of the seas, only on a vastly greater scale. The sun is supposed to act in two ways; first, by its attraction, or by its gravitation, and then by its heat. The attraction of the sun on a solid body surrounded by a nebulous or fluid atmosphere would necessarily change its figure from the spherical form to an oblong or ellipsoidal form, in case the exterior or disturbing force is very small in comparison with the central gravity of the body, and its diameter very small in comparison with the distance of the source of the disturbance. In this way the attraction of the sun and moon produces the ebbing and flowing of the tides of the sea, and of the atmosphere. In the case of a comet — the mass being small, and the volume extended — it is evident that the solar attraction can, at a certain distance, be much greater than that which exists between the different parts of the comet; and also, that this attraction will be very different in various parts of its volume. In order, therefore, to determine the figure of a comet at a given distance from the sun, it is only necessary to find the form which a fluid mass will assume, when the attraction of the sun is comparable with or even greater than the gravitation between its parts, its volume being considered as very great in comparison with its distance from him. Under these conditions the law of equilibrium of a moving mass has been determined, whose physical constitution is supposed to be a nucleus of solid matter surrounded by concentric layers of an elastic atmosphere, with a density continually decreasing from the centre outward. It should be remarked, how-

ever, that no direct solution of the question thus
presented has been satisfactorily accomplished; and
it is simply contended that, as a necessary consequence, the densest portion of the comet, supposed
to be a fluid mass, would not remain at the centre,
but would, on the contrary, approach the sun; and
that, in the case of a comet constituted chiefly of an
extremely attenuated vapor, its form would be
analogous to that of a column of expansible atmosphere, having its base supported by the attracting
body, the denser layers being always nearest the
base. In this way it is supposed that the tail of a
comet may be formed, resting in space, isolated and
extended, and always sensibly in a direction opposite to the sun. The case of a solid body surrounded
by a nebulous atmosphere is considered as the limit
of a minimum change of form, and that of an
atmospheric column as the maximum limit. Between these an infinity of varied forms may be
imagined; and hence this theory has been urged as
wholly sufficient to explain all the phenomena which
have been observed in the formation of the tails of
comets. The sudden and anomalous changes in
the vicinity of the nucleus as the comet approaches
the sun, are attributed to the influence of the solar
heat, which must necessarily produce an enormous
dilatation of its mass. It is further contended that
the expansion of the comet by heat, and its subse-

26 *   U

quent condensation or contraction by cold, in connection with the operation of the solar attraction,
as just explained, are sufficient to account for every
phenomenon connected with the structure of comets
which has hitherto been observed; and meteoro-
logical phenomena on our earth are adduced as
beautiful examples of similar effects. This theory
has been urged with great force of reasoning, and
has, indeed, been received with considerable favor;
yet it must be admitted that a detailed examination
of the recorded phenomena presented by some of
the largest comets, shows conclusively that it is
inadequate to their complete explanation.

A favorite theory of Newton was that the molecules
of the circumambient ether, overheated by means
of the comet, which absorbs the solar rays, and
transmits the heat to the ether, would become lighter
than its exterior layers, and would, therefore, flow
backward in a direction opposite to the sun, carrying with them a part of the molecules of the comet.
In this way it is imagined that the phenomena of
the tails of comets may be explained without resort-
ing to a direct repulsive force, thus avoiding completely whatever may seem to complicate and con-
fuse the beautiful conception of universal gravitation, by the introduction of an opposing force. This
theory of the action of the radiant heat of the comet
on the ether which surrounds it, and that, previously

stated, which connects the formation of the tails of comets with the combined action of the attraction of the sun, and the solar heat acting directly on the molecules of the comet, are those which have been classed as belonging to what may be termed apparent repulsion.

Kepler and others have attributed to the light of the sun a repulsive action, sufficient to drive off the lighter portions of a comet in the form of a tail. This is the case of a direct repulsive force, operating like gravity, only in a contrary direction. With this idea of repulsion as a basis, and omitting all considerations in respect to the character of light— as not being essentially of any account, in the con-clusions which may be arrived at — Faye has endeavored to develop a complete theory of the figure and physical constitution of comets. Before proceeding, however, to give an explanation of this theory, it may be well to state that Bessel and Olbers have attributed the formation of the tails and envelopes of comets to the operation of a repulsive force emanating chiefly from the sun, and being, like magnetism and electricity, a polar force. Olbers supposed the repulsive agent to be electricity, and imagined that it acted so as to produce a repulsive force in the sun, a similar force in the comet itself, and a specific action of the sun such as may be supposed to be exhibited in the simultaneous existence of several distinct tails. Bessel subjected the theory of Olbers to a rigorous mathematical calculation, introducing, however, such modifications as the exigencies of the case demanded. He supposed the action of the sun on a comet to be exercised in two different ways, namely, by a general action equal for every part, and by a differential action, strongest at those points nearest the sun, and most feeble at those parts most remote from him. This he conceived to be a necessary consequence, no matter what the nature of the force exercised may be. Thus, for example, in the case of the force of gravitation, there is a general action which produces the motion of the centre of gravity, and a differential action which causes a motion of rotation; the latter being always feebler than the former. He, therefore, supposes that the comet experiences at first the general action of the sun, and that, under this action, it emits from every part of its nucleus particles of matter whose polarity is *negative* with reference to the sun. On approaching its perihelion, the comet will experience, finally, the second mode of action, and this he imagines to give to it the two poles, the *positive* pole, with reference to the sun, being on the side nearest that luminary. At this point an emission of nebulous matter will commence; and, since the polarity is *positive*, this emission will be directed toward the sun. In passing through the nebulous atmosphere which is already filled with *negative* particles, and which is continually forming anew, the nebulous matter emitted will lose, by degrees, its primitive polarity, and even assume, finally, the opposite polarity, before it reaches the limit of this atmosphere. This emission from the nucleus will now be repelled by the sun, and driven back to form the tail. This is a general statement of Bessel's theory as modified from that of Olbers.

The phenomenon of the regular formation of luminous sectors, of which the great comet of 1811 furnished so beautiful an example, led Olbers to conclude, without hesitation, that the hypothesis made by Kepler and Euler, which we have already stated, was necessarily untenable, since the solar repulsion could not produce an emission directed from the nucleus itself toward the sun. It was for this reason that he considered it necessary to endow both the comet and the sun with repulsive forces, which he attributed to a common cause, namely, to the electricity which might be supposed to be developed by the approach of the comet to the sun. The periodic, or nearly periodic fluctuations of the luminous sectors exhibited in the case of Halley's comet, at its last return, led Bessel to assume the existence of a polar force developed in the comet under the influence of the sun, and in this way to

attempt to reconcile all the varied phenomena
which have been observed. Faye, adopting the
fundamental idea of a repulsive force, the existence
of which he considers to be conclusively established,
attempts to explain not only the figure of the comets,
but also the acceleration of their motion, which we
have stated to be due to the resistance of an ethereal
fluid pervading all space. He attributes to the
force of repulsion a mode of propagation, not instantaneous like that of gravitation, but in regular
succession, like the undulations of the ether in producing light. He supposes the repulsive force to
be resolved into two component parts, one acting
in the direction of the line joining the centre of the
sun with that of the comet, and the other in the
direction of a line tangent to the orbit of the comet
at the point where it is situated. The first component force may be supposed to determine the
figure of the comets, including the formation of
the tail, while the second may act to produce the
acceleration of their motions. The repulsive force
being supposed to be propagated by a series of successive undulations, it follows that it will not depend
directly on the mass of the sun, but on the extent
of his surface and its physical condition. The intensity of its action on any body whatever will vary
inversely as the square of its distance. In this way,
Faye considers it possible to give a most simple and

natural explanation of the phenomena which led to the hypotheses of Olbers and Bessel, thus avoiding a metaphysical objection, which he contends may be justly urged against the existence of a force such as these astronomers have assumed. The objection referred to is that these hypotheses attribute simultaneously to the same matter a repulsive force and an attractive force of the same nature, the same direction, and the same law, with the exception only of the specific and limited operation of the former, which distinguishes it from the universality of the latter.

In further confirmation of his theory, Faye adduces the fact that the theories of Olbers and Bessel seem to present almost insuperable difficulties in the explanation of the phenomena, which completely vanish when we adopt the simple and unique theory of Kepler, which he has adopted as the basis of his own. He regards the modifications of the original theory of Olbers introduced by Bessel as extremely complicated, in which it is supposed that, under the influence of the sun, the nucleus acquires polarity, and emits, in the direction of the sun, particles which are negatively electrified, while, at the same time, the sun exercises, on this part of the comet, a positive action. He regards it as an important objection to Bessel's theory, that, in order to explain the manner in which these particles or molecules of nebulous matter, notwithstanding their negative polarity, cease to be attracted by the sun, and finally become subject to an energetic repulsion on the part of the sun, which drives them back from him, it becomes necessary to suppose that by virtue of a general action of the sun — while the comet is yet far distant, and before the development of the polarity — the part of the comet from which the emission takes place has been positively polarized like the sun. He also cites the fact that Bessel admitted that the molecules which were thus differently polarized will become neutralized, and that the particles emitted by the nucleus finally lose their negative polarity, and assume a positive one, even before they are very distant from the nucleus, and ultimately become subject to the repulsive influence of the sun, being driven back to form the tail. The intensity of this force of repulsion will, it is urged, vary with the nature of these molecules. In the case of Halley's comet, at its last return, its molecules were repelled by the sun by a force nearly twice as great as that of attraction; and those of Donati's comet (1858) by a force whose intensity was a little more than one-third that of gravity in the case of the principal tail, and by a force nearly seventeen times greater in the case of the secondary train. The difficulty which Faye thinks very important, is to explain why the sun should exercise on the different particles of the comet two separate and distinct repulsive forces of such unequal intensity. This, however, has been explained by supposing that different parts of the comet had a vastly different specific gravity, being actually lighter than the surrounding ether, and that these particles are thus repelled through the ether with very different velocities — which was essentially the idea of Newton.

Soon after the appearance of the great comet of 1843, Professor Norton proposed a theory of the formation of the tails of comets which in some respects resembled the theories of Bessel and Olbers. He was led to the final hypothesis which he formed by such considerations as the following, having been entirely ignorant of the fact that in some points of his theory he had been anticipated by others. 1st. The general direction and situation of the tail with respect to the sun, show that this luminary is concerned, directly or indirectly, in its formation, while the changes which take place in the dimensions and figure of the comet, both in approaching its perihelion, and in receding from it, lead to the same inference. 2d. Since the tail lies in the direction very nearly of a line drawn from the centre of the sun through that of the comet, the particles of matter of which it is composed must have been driven off from the head by some force exerted in a

27

direction from the sun. 3d. This force cannot emanate from the nucleus, for such a force would expel the nebulous matter surrounding the nucleus in all directions, instead of one direction only. This objection, however, might be removed by supposing that a repulsive action is exerted by the nucleus only from that side which is most remote from the sun; yet it may be perceived that such a force as this would not give to the tail the form and direction which observations indicate, when it is considered that the tail has its origin in the nebulosity at the side of the nucleus. It is thus evident that no one force, having its origin in the head of the comet, can be conceived of which would be adequate to the production of the tail. There is, however, a possibility that a repulsive action of the nucleus on the matter of the nebulosity may be combined with some other force foreign to the comet, as an auxiliary cause in producing the phenomena of the tail; — its effects on the side toward the sun being contracted, at a definite distance from the nucleus, by this latter force. 4th. It seems to follow from what has just been stated, as a necessary consequence, that the matter of the tail is driven off from the head by some force foreign to the comet, and taking effect from the sun outward. 5th. This force, whatever its nature may be, must be supposed to extend far beyond the orbit of the earth; and although nothing can be predicated with certainty respecting the law of its variation, yet it is at least probable that, like all known central forces, it varies inversely as the square of the distance.

From such considerations as these Norton was led to the conclusion that there existed a repulsive force in the sun, and that although it might operate so as to drive off the nebulous matter to greater and greater distances, without destroying the connection of the parts—the head and tail thus revolving as one connected mass—or, so as continually to detach portions of the nebulosity, and repel them to an indefinite distance into free space; yet that it was most probable that the latter mode of action was the one which actually exists. After a long train of reasoning based on the phenomena of the comets which had hitherto appeared—but which it will be unnecessary to give in this connection—he arrived finally at the conclusion that the tail of a comet is made up of particles of matter continually flowing away, at a very rapid rate, from the head into free space, and that at any one instant we see the collection of all the particles that have been emitted during a certain previous interval. According to this hypothesis, at the end of any such interval we are looking at an entirely new tail. The particles may be supposed to be detached from the outer portions of the nebulosity of the comet, by the repulsive force of the sun; in which case they would fly off in the directions of the lines diverging from the sun, along which the force acts, but would be made by the attraction of the nucleus to pursue paths slightly deviating from these lines and concave toward the axis of the tail; or, as Bessel and Olbers have supposed, they may undergo some modification by the action of the sun, by reason of which they are first repelled outward from the nucleus, and then driven away from the sun, into the depths of space, by virtue of his superior repulsion. According to the latter supposition, they acquire an initial velocity in leaving the nucleus, and subsequently, under the action of the sun's repulsive force, they will move off in hyperbolas, having the sun in their remote focus, and concave toward the axis of the tail.

By means of this theory Norton was enabled to explain the curved form of the tail, and the general fact of its deviation from the position of direct opposition to the sun; and also to indicate the causes in operation to produce the various degrees of curvature, and various positions of the tail, noticed in the case of different comets, and in different positions of the same comet in its orbit. The appearance of Donati's comet in 1858 induced him to modify his theory in some particulars, and to reduce it to a complete and systematic form. The phenomena of successive envelopes presented by this comet to the most remarkable degree, led him to conclude that the nebulous envelope of the head of a comet was probably in the same dynamical condition with the tail; that the cometic matter was, in the first instance, expelled from the nucleus on the side toward the sun, and then driven back to form the tail, by virtue of the repulsive force emanating from the sun; and that this process went on continuously; the nebulous matter rising perpetually, like a gushing fountain of light, and the luminous jets bending back, in graceful curves, before the repelling energy of the sun. He, therefore, regarded the nebulous envelope of the head, and the tail of a comet, as simply different parts of the same flowing stream; and imagined that the cometary matter was, in the first place, urged away from the nucleus by an incessant force of repulsion exerted by its mass, and that this force and the repulsive action of the sun, like all radiant actions, varied inversely as the square of the distance. Other suppositions were also made — as that the expelling action was an instantaneous repulsive force; and again, that the nebulous matter was first detached by a superficial force of projection, and subsequently acted on by the repulsion of the mass of the nucleus. By means of this theory, Professor Norton was enabled to explain all the phenomena presented by the great comet of 1858,

27 *

and also to derive an approximate value of the density
of the matter composing the nucleus of the comet.
He found that the nucleus of this comet was of from
five to six times the density of water, or about the
same as the mean density of the earth. He performed similar calculations in the case of four other
comets, and found that their density is from four to
fourteen times that of water. These results may
seem to be at variance with the generally received
opinion in regard to the physical constitution of
these bodies; and it must be admitted that they are
at most but rude approximations; — yet, he considered them as affording sure indications that the
telescopic nucleus of a bright comet is not entirely
gaseous, or nebulous, but that it is either liquid or
solid, or composed of both liquid and solid matter,
like the earth.

He accounted for the rise and gradual recess of a
succession of envelopes from the nucleus, by supposing that the evolution of cometic matter first
begins at the polar regions of the nucleus, and
gradually extends toward the equatorial parts, the
comet being supposed to rotate about an axis perpendicular to the plane of its orbit. Under these
circumstances, if the comet is observed under a
large angle to the plane of its orbit, as was the case
with Donati's comet, after the perihelion passage,
the outline of the visible envelope would steadily

enlarge, as the process of evolution was propagated
further and further from the polar regions of the
nucleus.   In this explanation, he supposes that the
envelopes are not spherical in form, but that each
approximates in its form to that of a flat semicircular disc, of a definite thickness.   The dark stripe
extending from the nucleus to a considerable distance in the tail, which we have already noticed in
the description of Donati's comet, is supposed to
show conclusively that the envelopes from which
the tail proceeded were much flattened in the
direction of the line of sight; or approximately in
that direction.

Such are the principal theories which have been
devised, in order to explain all the anomalous
appearances presented by the comets; and a careful
consideration of these different hypotheses in connection with observed facts, will make it evident
that there exists in nature a general force of repulsion, exerted by all masses, and operating under
certain circumstances upon matter in an extremely
attenuated state.   The idea of repulsion being thus
established on a firm basis, as a physical fact, it is
only necessary to determine which of the various
theories of the mode of operation of the repulsive
force will best explain all the varied phenomena of
comets in general.   The hypotheses of Bessel, Olbers,
and Norton, are essentially the same, and differ only

in certain supposed peculiarities of the force of repulsion, and in its mode of operation. Bessel was the first to apply a vigorous mathematical analysis in the determination of the effects of such a force as is found to exist. The modifications which he found it necessary to introduce into Olbers' theory, in order to explain the phenomena observed in the case of Halley's comet, at its last return, have been objected to as rendering the results contradictory when an attempt is made to derive the intensity of the forces in operation. The appearance of the great comet of 1858 under circumstances peculiarly favorable to the final determination of the theory of the formation of the tail by means of a repulsive force common to all bodies, required simply that the various theories advanced should be subjected to the test of exact mathematical calculation. This was performed by Professor Pierce in the case of Olbers' hypothesis, and the results obtained were in such strict accordance with the observations, that this theory is now regarded as being established on the surest possible foundation.

There is an inference which may be drawn from an examination of the theory of the operation of a repulsive force in the formation of the tail of a comet, which is, that some portions, at least, of the nebulous matter will be detached entirely from the comet, and driven off into space. Under this view it may be supposed that the comets are gradually wasting away, and will finally be extinguished. Observations, however, do not by any means render it certain that these bodies are thus affected, and the necessity of such a supposition may perhaps be obviated by regarding the ethereal fluid which pervades all space as of sufficient density to afford a sensible resistance to the flow of the nebulous matter from the head of the comet, thus limiting the extent of the tail. There are, indeed, questions here presented which we are unable to answer satisfactorily, and which require a long series of observations of the successive returns of a comet, for their elucidation. It may be that the comets are thus wasting away, but that in their wanderings through endless space they casually assimilate new cometic matter from the waste of other comets, and thus undergo a series of perpetual changes. It may also happen that two different comets may approach each other near enough to cause their separate masses to coalesce, thus furnishing abundant material for future losses. These and similar questions will undoubtedly be finally solved, though the observations hitherto recorded fail to give a satisfactory answer.

Another consideration which may appear in this connection, is, whether the tail of a comet, being composed of attenuated nebulous matter, can offer

any sensible resistance to the regular motion of the
centre of gravity of the nucleus. In observing these
bodies, their places on the celestial vault are determined under the supposition that the point of
greatest brilliancy corresponds to the centre of
gravity of the comet. The elements of the orbit
are computed under this hypothesis, and it might
be expected that the predicted places would not be
found to correspond with the actual observations.
In the case of all the comets, this is to a greater or
less extent true, but the deviations have usually
been attributed to the necessary uncertainty of the
observations. In the case, however, of Donati's
comet, it is found that it is absolutely impossible to
represent the entire series of observations by a
purely elliptic orbit, and that the discrepancies
between computation and observation are such as
to indicate beyond a doubt that the motion of the
nucleus in its orbit is sensibly affected by the tail,
or by the joint action of this and the repulsive forces.

The question may also arise as to the nature or
cause of the repulsive force whose existence is thus
so clearly established. The near approach of a
comet to the sun, under an ever-varying temperature, would necessarily produce, when near the sun,
the most intense electrical excitement; and we may
thus infer that the force partakes in some degree of
the nature of electrical action. The existence of

an ethereal fluid pervading space is beyond all
doubt, and it is certainly known that light is produced by certain vibrations or pulsations, through
this ether, an effect which is imparted, to a greater
or less extent, by all material bodies. The development of electric currents, as well by magnetic as
by electrical action, the similarity in their mode of
action in a great variety of circumstances, and
especially the production of the spark from a magnet,
the ignition of metallic wires, and chemical decom-
position, all tend to show that magnetism and electricity are to be regarded as identical. The evolution of light and heat during the passage of the
electric fluid, the development of electricity by heat,
the influence of heat on magnetic bodies, and that
of light on the vibration of the compass, show
decided indications of some common and invisible
bond uniting these mysterious but powerful agents
of nature. The fact that these agencies operate
each with an intensity varying in accordance with
the same law, is a further proof of their common
origin. The solar spots, it may likewise be remarked, are found to be connected intimately with
terrestrial magnetism, and in view of all these facts,
it seems reasonable to conclude that light, heat, and
electricity, are all produced by successive undulations in the ethereal fluid—just as sound is caused
by vibrations communicated to our fluid atmosphere,

—yet under different circumstances with respect to
the operation of the cause.

Having thus considered in detail the various
theories of the physical constitution of these wandering bodies, it may not be improper to add a few
remarks in reference to their probable origin. Laplace, following the speculations of Sir William
Herschel, applied the nebular theory of that astronomer to the formation of the solar system, com-
prehending the comets as well as the planets and
their satellites. He supposes that in the primeval
condition of the solar system, the sun revolved on
his axis, surrounded by an atmosphere which, on
account of an excessive heat, extended far beyond
the orbit of the remotest planets. The heat gradually diminished, and as the solar atmosphere contracted by cooling, the rapidity of his rotation became increased, and the exterior portions of the
nebula by which he was surrounded, would be
detached from the rest, the central attraction being
no longer able to overcome the increased centrifugal
force. This ring of nebulous matter would, from
various causes, be broken into fragments forming
a planet and satellites, or, which is more probable,
the fragments would coalesce into one nebular mass,
which would revolve around the sun in an orbit
nearly, if not quite circular, lying in a plane nearly
coincident with the plane of the equator of the

central body, and revolving in its orbit in the same direction in which the central globe rotates on its axis. The same process would again take place when the central mass had again parted with a sufficient quantity of heat, and thus the successive planets may be supposed to have been formed in a state of vapor, the process continuing until the cohesion of the particles of the central mass were finally able to resist any further change. These planets, in their turn, having each its motion of rotation, would, as they became gradually cooled and condensed, produce satellites in the same manner, and by the operation of the same laws, by virtue of which they were themselves formed from the nebulous matter surrounding the sun. It may be perceived also that the motions of the satellites thus produced, and the motions of rotation of the planets, must be in the same direction. Thus may this strange and apparently fanciful theory be made to account for the most remarkable circumstances in the structure of the solar system. The motions of the planets in the same relative direction, and almost in the same plane; the motions of the satellites in the same direction as those of the planets; the small eccentricity of the orbits of the planets, upon which condition, together with others to be noticed in a subsequent connection, the stability of the system depends; and the position of the source

28

of light and heat in the centre of the system, are all satisfactory by this hypothesis. This is the *nebular hypothesis* of Laplace, and although it was originally advanced as simply a theory of the development of the solar system from one primitive mass of nebulous matter, yet the facility with which it explains various phenomena connected with the structure of the system has induced many to regard it as strictly true.

Laplace supposed that in the very beginning of the creation of our system, the sun consisted of a diffused and extremely attenuated nebulous matter, extending over a vast extent of space. This nebulous matter may have been so exceedingly attenuated that its existence was barely a fact. As a confirmation of this view of the case, he sought an analogy in those nebulæ which are now seen here and there throughout the entire heavens, so feeble and diffused as to present the appearance simply of an ill-defined mass of vapor when seen through the most powerful telescopes. He imagined that in the vast assemblage of irresolvable nebulæ — by which is meant, those which the most powerful telescopes fail to reduce to a group of minute stellar points of light, or individual stars — there were examples of similar processes of development now going on in various parts of the universe. In the first place we see the nebulous matter dispersed in patches in the different

parts of the sky, while in some of these it is supposed that traces are exhibited of this matter feebly
condensed around one or more faint nuclei. In
other cases these nuclei are brighter in proportion
to the surrounding nebulosity, and further on, in
the apparent progress of development, there are
supposed examples of the condensation of the atmosphere of each nucleus, thus exhibiting a collection
or aggregation of nebulous stars, formed by brilliant
nuclei very near each other, and each surrounded
by an atmosphere. Still further on in the process
of development the nebulous matter, by condensing
uniformly, is supposed to form nebulous systems
which are called planetary; and, finally, a much
greater degree of condensation transforms all these
nebulous systems into stars. If this hypothesis be
true, then will all those nebulæ which are now
irresolvable, even in the most powerful telescopes,
be ultimately transformed into stars, while it may
be inferred, from a similar course of reasoning, that
the anterior or primeval condition of the stars which
now exist, was that of nebulous matter highly
attenuated and distributed throughout space. This
is the most general view of the nebular hypothesis,
and granting what has been stated, we may conclude
that at this very moment creation is going on in
distant parts of the universe, and will continue for
ever. It must not be objected that there is anything

in this view of the creation which militates against revealed religion. To supply the primitive matter, and endow it with such properties as must have been required to effect the various transformations from inanimated existence to animated existence of the highest order, must indicate the power and presence of the Almighty Creator. It may also be objected that in all such speculations we enter the confines of the unknown; yet the anxious search for truth which characterizes the human mind must be considered a sufficient excuse for thus wandering into the realms of the invisible and uncertain.

According to the nebular hypothesis of the formation of new worlds from original and primitive masses of matter, the comets have their origin in portions of the nebulous matter occupying positions intermediate between two or more great centres. These masses of attenuated matter are supposed to be held in a state of equilibrium, until, finally, the attraction of some one centre predominates, and the uncondensed filmy mass begins to descend slowly toward its controlling centre. The fact that the comets come into our system in all possible directions, and in orbits which conduct them far beyond the limits of our system, is in itself conclusive evidence to show that they wander here and there among the various systems constituting the universe. This would naturally be supposed to be the case if they were formed in accordance with the nebular hypothesis, their uncondensed condition being due to the fact that the feeble gravitation of their parts toward the centre of the mass is inadequate to contract them into a more solid form; and also that, in passing near the suns which they visit, they are necessarily converted by the intense heat into an extremely vaporous state. It may be that in the course of ages the effects of the cold in the remote parts of their orbits would solidify them to such an extent that in future they might retain, to a limited extent, a solid form; the exterior portions, however, of the compact nucleus, in their successive revolutions, going through alternate changes of solidification and evaporation. We may remark also that the idea of attributing the origin of comets to the nebular hypothesis, does not in the least interfere with or modify the theory of the formation of their tails already given.

We have now given explanations of the principal phenomena presented by the comets directly, which the researches of astronomers have indicated, and thus far our task is accomplished. There is, however, one other consideration, which the study of the motions of these wonderful bodies has presented, and to which we shall next advert; namely, the existence of a medium of resistance pervading space, and sensibly opposing the motions of the heavenly

28 *

bodies, at least those which are known as cometary worlds. We cannot, however, conclude the consideration of this branch of our subject without remarking further, that the phenomena of the comets are the most varied, the most beautiful, and, at the same time, fantastical, which the heavens present to our view. The peculiar forms of their orbits, the suddenness of their appearance, and the development of their physical changes as they approach the sun and again recede into illimitable space, all tend to afford a sublime conception of the grandeur of God's creation. Even if we ever feel inclined to doubt in regard to the endless extent of the universe, the circumstances which are so undeniably proved to be connected with their motions, are sufficient to convince us of the reality of infinity. Following, therefore, these mysterious bodies in our imagination in their endless wanderings through space, we may perhaps be prepared to realize, to some extent, the beautiful yet true conception of that celebrated dream of Richter:

"God called up from dreams a man into the vestibule of heaven, saying, 'Come thou hither, and see the glory of my house.' And to the servants that stood around his throne he said, 'Take him, and undress him from his robes of flesh: cleanse his vision, and put a new breath into his nostrils; only touch not with any change his human heart —

the heart that weeps and trembles.' It was done:
and, with a mighty angel for his guide, the man
stood ready for his infinite voyage; and from the
terraces of heaven, without sound or farewell, at
once they wheeled away into endless space. Sometimes with solemn flight of angel wing they fled
through Zaarrahs of darkness, through wildernesses
of death, that divided the worlds of life; sometimes
they swept over frontiers that were quickening
under prophetic motions from God. Then, from a
distance that is counted only in heaven, light dawned
for a time through a sleepy film; by unutterable
pace the light swept to *them*, they by unutterable
pace to the light. In a moment the rushing of
planets was upon them: in a moment the blazing
of suns was around them.

" Then came eternities of twilight, that revealed,
but were not revealed. On the right hand and on
the left towered mighty constellations, that by self-
repetition and answers from afar, that by counter-
positions, built up triumphal gates, whose architraves, whose arch-ways — horizontal, upright —
rested, rose — at altitude by spans — that seemed
ghostly from infinitude. Without measure were
the architraves, past number were the arch-ways,
beyond memory the gates. Within were stairs that
scaled the eternities below; above was below —
below was above, to the man stripped of gravitating

body: depth was swallowed up in height insurmountable, height was swallowed up in depth unfathomable. Suddenly, as thus they rode from infinite to infinite, suddenly, as thus they tilted over abysmal worlds, a mighty cry arose — that systems more mysterious, that worlds more billowy — other heights and other depths—were coming; were nearing; were at hand.

"Then the man sighed, and stopped, shuddered, and wept. His overladened heart uttered itself in tears; and he said — 'Angel, I will go no further. For the spirit of man acheth with this infinity. Insufferable is the glory of God. Let me lie down in the grave, and hide me from the persecution of the infinite; for end, I see, there is none.' And from all the listening stars that shone around issued a choral voice, 'The man speaks truly: end there is none that ever yet we heard of.' 'End is there none?' the angel solemnly demanded: 'Is there indeed no end? — and is this the sorrow that kills you?' But no voice answered that he might answer himself. Then the angel threw up his glorious hands toward the heaven of heavens, saying, 'End is there none to the universe of God. Lo! also there is no beginning.'"

CHAPTER V.

THE question of a *plenum* and a *vacuum* has engaged the attention of philosophers from the earliest
ages of the world.   In their speculations respecting
the constitution of the universe, it often became a
matter of doubt whether the celestial and terrestrial
spaces are absolutely full, or whether there are, between and among the material parts of the universe,
empty spaces free from all matter.   The general
opinion, however, seems to have been that the
regions around us, beyond our atmosphere, and to
an indefinite extent, are supplied with a rare invisible medium, of unknown composition and character,
in which all bodies are compelled to move.   In the
astronomy of the Brahmins of India, which dates
its origin long before the commencement of the
Christian era, the existence of such a medium is
assumed, though the stars were by them supposed
to move in it like fishes in water.   They regarded
it as a celestial element, pure and impalpable, and

offering no resistance to the motions of the heavenly
bodies. Indeed, almost every ancient astronomer
·held similar opinions. Kepler, notwithstanding his
profound knowledge of astronomy, in seeking the
origin of comets, supposed them to be native inhabitants of this fluid medium, as the fishes are of
the waters of the earth; and that they were thus
created to inhabit the immense spaces of the universe. The sombre and bloody appearance which
the sun sometimes exhibits, he attributed to a
coagulation of the ethereal fluid, and the cessation
of these appearances he supposed to be due to the
collection of the denser portions of the ether in the
form of comets.

The theory of an ethereal medium throughout all
space remained thus until the development of a
system of the world by Descartes. He supposed
that matter possessed only the properties of extension, impenetrability, and inertia; that it filled
all space; and that it was, in its parts, both great
and small, endued with motion in an infinity of
directions. He imagined that the matter would
thus be formed into innumerable vortices, differing
not only in extent, but also in velocity and density;
the more attenuated parts constituting the real
vortex, in which the denser bodies float, and by
which they are pressed, though not equally, on all
sides. He therefore regarded the universe as con-

sisting of innumerable vortices, which limit and circumscribe one another, and supposed that the earth and planets are bodies carried round in the great vortex of the solar system. He supposed also, that, by the pressure of the attenuated matter which he had assumed to fill all space, and which circulates with great rapidity and great centrifugal force, the denser bodies, which have less rapidity and less centrifugal force, are forced toward the sun, the centre of the vortex. In like manner, each planet was itself regarded as the centre of a small vortex, by the operation of which all the phenomena of gravity would be produced. This is the Cartesian system of the world, a theory which was very generally received for nearly half a century from the date of its promulgation; and it is evident that the existence of an ethereal fluid pervading all space is here adopted as the basis of the theory. The glaring defects, however, of the doctrine of Descartes, were conclusively established by the subsequent discoveries of Newton and Kepler. It was shown by the former, that the planetary and cometary bodies revolve around the sun by virtue alone of the force of gravitation, and the original force of projection imparted to them at the time of their formation. That this theory was correct, the complete verification of Kepler's laws exhibited the most incontestable evidence, while the subsequent development of the great structure which modern astronomy forms, has completely verified and generalized it in every particular.

The Newtonian theory of gravitation was conceived by some to require that the motions of the heavenly bodies should be performed, according to mechanical laws, in space which was absolutely a vacuum; and hence, that the doctrine of an ethereal medium of any conceivable density was inadmissible. This, however, is by no means a necessary consequence of the law of universal gravitation, and was not so considered by Newton himself. He believed firmly in the existence of a subtile ethereal fluid pervading all space, and even the internal parts of all material bodies. He supposed that possibly it was this medium which furnished the means for the operation of the gravitating force. The ethereal fluid was conceived of as a sort of spirit, by the force and action of which the particles of bodies mutually attract each other at near distances, and cohere if contiguous. He imagined that possibly this medium is much rarer within the dense bodies of the sun, stars, planets, and comets, than in the empty spaces between them; and that in passing from them to great distances, it became gradually denser and denser, thus, it may be, causing the gravity of these bodies toward each other, and of their denser parts toward a centre; every body tend-

ing continually to move from the denser portions
of the medium toward the rarer. These hypotheses
in regard to the existence, nature, and effect of the
ethereal fluid, were not, however, advanced by Newton as any part of his theory of universal gravitation, but simply as a possible method of explaining
the operation of the gravitating force. This was
all that it was necessary for him to maintain, since
his theory of the system of the world was founded
exclusively on observations, and had nothing to do
with abstract possibilities and metaphysical neces-
sities.

The motions of the planets computed strictly in
accordance with the law of gravitation, and under
the supposition that they move in an absolute
vacuum, or, at least, in a space void of any resisting
fluid, were found to accord so closely with observation, that it was for some time doubted whether any
such medium as Newton and others had suspected,
really existed. The phenomena of the planetary
motions were so accurately predicted by the Newtonian theory of gravitation, that it became evident
either that there was no such fluid, or that it was so
extremely attenuated or rarefied, that no phenomenon which had hitherto been observed was capable
of exhibiting its effects. In order, however, to test
this question thoroughly, the Academy of Sciences,
of Paris, in 1762. offered a prize for the solution of

the following definite problem: *"Do the planets re-volve in a medium of which the resistance produces a sensible effect upon their movements?"* The prize was awarded to Bossut, who found, by calculation, that the effect of any such resistance, offered to the planets, would be to diminish the greater axis of their orbits, and, consequently, to shorten their periods of revolution. The moon was known to revolve in periods which were becoming shorter and shorter, and he concluded that the observed accelera-tion was due to the resistance of the ether. He applied the same reasoning in the case of the planets, and although no such acceleration could be detected, yet he conjectured that in the course of time similar effects would be exhibited. It should be remarked, however, that up to the present time no acceleration of the planets, due to action of a resisting medium, has been detected; and the acceleration of the moon's mean motion is explained thoroughly and satisfactorily by the changes which the perturba-tions by the sun and planets are producing in the eccentricity of the earth's orbit—thus showing con-clusively, that so far as the motions of the planets and their satellites are concerned, we have no evi-dence of the existence of a resisting ethereal fluid in space.

We have already stated the fact that light is cer-tainly known to be produced by successive undula-

tions or pulsations in an extremely attenuated and
elastic ethereal fluid, which pervades all space; and
that heat and electricity may with great probability
be attributed to the operation of the same cause,
only under different circumstances. Such being the
case, it seems evident that this ethereal fluid will
oppose itself to the motions of the heavenly bodies
in their orbits, and that its effect on the motions of
the comets would be first exhibited, since the extremely vaporous matter of which they are composed would be most readily operated upon by the
resisting medium. If, therefore, it should be found
that the comets which return regularly and at short
intervals, actually exhibit the accelerated motion
which would result from the operation of such a
medium of resistance, it may be regarded as conclusively established, that the undulatory theory of
light is a physical fact, and that there exists throughout this illimitable universe a subtile ethereal fluid.
The only objection which can be urged against this
assumption is that the resisting medium must necessarily be composed of ponderable matter, while the
ethereal fluid, which serves for the transmission of
light, is supposed to be strictly imponderable. This
objection has indeed been urged in opposition to
the theory of a resisting medium, but it should be
remembered that the idea of imponderability is
essentially relative. The comets themselves, to a

limited extent, are imponderable, and when it is considered that the tails of these bodies are so extremely attenuated that they approach almost to the nature of the invisible ether, the identity of the resisting medium which opposes their motions, and the ethereal fluid which transmits light, is by no means doubtful. The phenomena of double refraction and polarization have demonstrated with unerring certainty that the undulatory theory of light, which presupposes the existence of a subtile ethereal fluid throughout space, is the only one which can be recognized in the present state of optical science; while other considerations, which have already been stated, tend to prove that there exists some common and yet inscrutable bond, which connects these powerful agents of nature. There can be no doubt, therefore, even if there were not additional phenomena, and of an entirely different character, tending to the same final consequence, but that the ethereal fluid exists in every part of the universe in which creation is already begun or ended, and that this subtile fluid must necessarily oppose itself to the free motion of all material bodies. There is nothing in this view of the system of the world which can be objected to on account of any assumed insufficiency of the data from which such conclusions have been derived, or, as being in contravention with the teachings of revelation.

The existence of a resisting medium in space may thus be supposed to be established on a purely theoretical basis, and it remains now to see whether there are any phenomena connected with the motions of any of the heavenly bodies, by means of which, even if no such ideas had been previously advanced, the same conclusion would have been derived. The planets do not, during the period in which they have been accurately observed, afford any indication of an opposing force, such as would result from a resisting medium. Their great weight, compared with the extent of their volume, would necessarily render them less liable to be sensibly affected by such a force of resistance. The case of the comets, however, is far different. These bodies are themselves composed, for the most part, of matter so attenuated, while the extent of their volume is often enormous, that the greatest resistance would be opposed to their motions; and, consequently, it is to these bodies that we are to look for an exemplification of the effects of a resisting medium. It has already been noticed in the case of Halley's comet that the effect of a resisting medium on its motions, from 1682 to 1759, was computed by Clairaut, and found to be so small as to be almost, if not entirely insensible. Such being the case, the effect of a resistance from the ethereal fluid was not considered in the computations which were made for its return

29 *

in 1835. It has also been noticed in the case of Encke's comet, and of Biela's comet, that their successive returns to their perihelia take place at intervals which are regularly becoming shorter and shorter. This diminution of the period of these comets has been attributed to the influence of a resisting medium, and the amount of the diminution of the period corresponds precisely with what might be expected to result from such a cause. In the case of Encke's comet the diminution of the period has been determined with the very greatest precision, the short period of its revolution affording great facilities for settling the question as to the existence of a medium of resistance. Encke has computed the planetary perturbations for the entire period which has elapsed since the comet first became known to astronomers, and finds that, after making due allowance for these disturbances, its period is shorter at each successive return. The amount of the diminution of its periodic time is very nearly eleven-hundredths of a day, or two hours, thirty-eight minutes, and twenty-four seconds, at each return. The same result has been obtained in the case of Biela's comet, but since the perturbations have not yet been computed for its successive returns with extreme accuracy, and, moreover, on account of the anomalous phenomenon of its having separated into two separate and distinct comets, the exact amount of the acceleration has not been ascertained. There can be no doubt, however, but that its period is becoming shorter at each successive return to its perihelion, and that the other comets of short period will eventually, after a sufficient number of revolutions, be found to exhibit precisely the same phenomenon. It may, therefore, be considered as well established that the motions of the comets indicate, beyond all doubt, the existence of a resisting medium within the limits included by the solar system, while the identity of this with the ethereal fluid which transmits light, being admitted, shows that the same medium of resistance is universal in the material universe.

The effect of a resisting medium on the motions of the bodies which revolve around the sun will be to continually accelerate their mean motion, and, consequently, shorten their period of revolution. The eccentricity of their orbits will gradually diminish, as the greater axis decreases, and, finally, they will one by one be plunged into the sun, after having described an indefinite number of circumvolutions of a kind of spiral curve, which gradually at first, but afterward rapidly brings them to the central body. The comets being most affected by the resistance will necessarily be the first to fall to the sun, and then in rapid succession the various planets belonging to our system; until finally all traces of the former existence of a beautiful and yet complicated system of comets and planets will be completely obliterated. This will be the necessary result of the resistance of an ethereal fluid in space, though it shall be accomplished only after the lapse of almost an indefinite number of ages. It may, therefore, on account of our preconceived idea of infinite duration, seem very improbable, if there be not other forces which can overcome the effect of a resisting medium, that it can really exist, notwithstanding the facts already presented. The conception of universal gravitation is so simple and yet so beautiful, that when we come to consider its legitimate results, in connection with the existence of a great central sun, around which the systems and compound systems of the universe revolve, the mind hesitates to assume the existence of any force either of repulsion or of resistance, which can limit the period of these revolutions and counter-revolutions. For example, supposing it possible to view our own solar system from some point in space beyond its limits, and that it is possible to witness both the revolutions of the planets and comets around the sun, and the revolution of the entire system around its distant centre, let us contemplate in imagination the scene which would be presented to our view. In the centre of the system would be seen the great central orb, or sun, and around him the planets, some with and some without satellites, each performing its appointed revolution. Here and there among these would be seen chaotic worlds, with streams of light flowing from them, some moving rapidly toward the central body with a constantly accelerated velocity, and others moving from it with a velocity decreasing just as in the other case it increased, while with all this gorgeous equipage of glittering worlds, the great sun itself and system would be seen to move slowly but uniformly around its centre. This also might be conceived of as revolving around a second great centre, and this again around a third, and so on until, finally, the mind rests on the contemplation of a great centre of centres of revolution, around which the entire universe is brought to an equipoise. This conception of the structure of the universe may perhaps be regarded as fanciful, and doubtless, to a greater or less extent, it is so; yet there is in it that which causes the reason at first to revolt against the assumption of the existence of any forces in nature which may ultimately destroy the present state of things, as contrary to the evident design of an intelligent and omnipotent Creator. To this may be traced the great opposition which the theory of the existence of a resisting medium in space has received, and which has led to various other hypotheses to explain the phenomena of the diminution of the periodic times of the comets.

Without attempting to consider the various theories which have been devised in order to avoid the assumption, or rather the confirmation, of the existence of an ethereal fluid pervading all space, which opposes itself to the motions of the heavenly bodies, let it suffice simply to state that all these are so incomplete and unsatisfactory, that, however much we may feel inclined to do otherwise, we are compelled finally to resort to the hypothesis of a resisting medium in order to explain the phenomena observed. This we are compelled to do in order to satisfy the results of observation, while, for reasons which have already been given, we must admit the existence of an ethereal fluid, whether sufficiently dense or not to oppose sensibly the motions of the heavenly bodies. It may be perceived, therefore, that the theory of a resisting medium presents the very greatest claims to our consideration, as the only one which can explain simultaneously several of the most important facts connected with the constitution of the material universe; and, consequently, before proceeding to investigate more fully its ultimate effect on our system, it may not be improper to notice briefly the provisions which the operation of the law of gravitation has afforded for securing the permanence of the system.

From what has previously been stated in regard to the mutual attractions of the heavenly bodies, it may be understood that, since all the planets are small compared with the sun, the derangement which they may produce on any one of their number will be very small in the course of a few revolutions. But since the same forces are constantly in operation, and must continue so, it may be easily conceived that in the course of ages the derangement of the planetary motions may accumulate, the orbits may change their form, and their mutual distances may be greatly increased or diminished. These changes might be supposed to continue without limit, and end finally, though perhaps at a date almost infinitely remote, in the total subversion and ruin of the system. Moreover, the careful observation of the planets for a long period of time has shown that changes are actually taking place in their relative motions, of such a nature as just referred to, which have been going on progressively since the very dawn of science. The eccentricity of the orbit of the earth has been decreasing for thousands of years, and, consequently, the mean motion of the moon has become more and more accelerated. The moon, therefore, perpetually approaches nearer and nearer the earth at each revolution, and should this change continue forever, it would eventually fall to our globe, an event which

would indeed be a dreadful calamity. The question
may thus arise, whether these and similar changes
will continue indefinitely, or whether, in the nature
of the planetary orbits, there exist conditions which
fix the limit of all these changes, and thus preserve
the stability of the system.

The problem which is thus presented is certainly
worthy of the highest exertions of the human mind,
while its solution is a task of no ordinary difficulty.
The problem is indeed one of such proportions that
no attempt was made to demonstrate either the
stability or instability of the solar system, until
near the close of the eighteenth century. It was
then shown by Lagrange and Laplace, that the
arrangements of the system are stable, that in the
course of time counter-changes will take place, so
that in the end the planetary orbits and motions
remain unchanged. The variations of the elements
of their orbits, caused by their mutual attractions
in accordance with the law of universal gravitation,
are periodical, and not indefinitely progressive. The
periods, however, in which these changes and
counter-changes are effected, are indeed enormous;
not less than thousands, and, in some instances,
millions of years; and hence it is that some of these
apparent derangements have been progressing in
the same direction since the very beginning of the
history of the world. The disturbances which will

ever exist will not be sufficient to modify or alter the adaptations of the system.

It was demonstrated by Laplace, that whatever be the masses of the planets, in consequence of the fact that they all move in the same direction in orbits of small eccentricity, and slightly inclined to each other — their secular inequalities or changes are periodical, and included within narrow limits; so that the planetary system will only oscillate about a mean state, and will never deviate from it, except by a small quantity. The orbits of the planets have been, and will always be, nearly circular. The ecliptic will never coincide with the equator, and the entire extent of the variation in its inclination cannot exceed three degrees. Thus will the planets, so far as gravity alone is concerned, continue to revolve forever in orbits, whose planes will rock slowly up and down, gradually contracting and expanding, but yet compelled to oscillate about a mean orbit whose form and position is fixed.

Such is the result of computations which are of the most positive character; and it is beyond all doubt that if the force of gravity alone operates on the bodies of the system, it will ever continue, so far as the effect of the planets may be considered, to preserve the stability of the system. By it these changes are produced, and by it the limits of the changes themselves are fixed. These limits can

30

never be passed, and after the lapse of perhaps millions of ages, the entire system will have resumed its primitive condition. But if now we introduce into the system an opposing force, such as would result from a resisting medium, the beautiful arrangement of the configurations of the planets which, by the operation of the gravitating force preserves and perpetuates the conditions of stability, will no longer exist. Unless, therefore, we can conceive of other forces in nature which can compensate for the effect of a resisting medium, we must regard it as inevitably decreed that all must end. This would be a necessary and unavoidable result, and its final consummation will depend simply on the character of the medium which exists. Thus, it has been found that Encke's comet would lose one-half of its present mean velocity in less than twenty-three thousand years. In a similar manner, it has been found, or rather conjectured, that if Jupiter were to lose a millionth part of his velocity in a million of years — which is regarded as far more than may be considered in any way probable — he would require seventy millions of years to lose one-thousandth of his present velocity, and a period seven hundred times as long to reduce it to one-half. These immense periods are almost sufficient to overwhelm the imagination — nevertheless it is certain that even if such changes are to be ultimately effected, the time which must elapse will be much greater. It may indeed be millions of millions of years before the earth's retardation will sensibly change the length of our year and the course of our seasons, yet, if there are no forces to counter-balance, even this exemplification of the smallness of the resistance does not in the least indicate its uncertainty or that of the fate which thus inevitably awaits our planet and its inhabitants, should they be permitted to exist for so long a time. The same effect will be produced throughout the system, and unless counter-changes shall be produced, it is indeed an appalling fact that all must end.

Such considerations as these have induced Who-well to indulge in the following reflections: "The vast periods which are brought under our conside-ration in tracing the effects of the resisting medium, harmonize with all that we learn of the constitution of the universe from other sources. Millions and millions of years are expressions that at first sight appear fitted only to overwhelm and confound all our powers of thought; and such numbers are no doubt beyond the limits of anything which we dis-tinctly conceive. But our powers of conception are suited rather to the wants and uses of common life, than to a complete survey of the universe. It is in no way unlikely that the whole duration of the solar system should be a period immeasurably great in our eyes, though demonstrably finite. Such enormous numbers have been brought under our notice by all the advances we have made in our knowledge of nature. The smallness of the objects detected by the microscope and of their parts;— the multitude of the stars which the best telescopes of modern times have discovered in the sky;—the duration assigned to the globe of the earth by geological investigation;—all these results require for their probable expression, numbers which, so far as we see, are on the same gigantic scale as the number of years in which the solar system will become entirely deranged. Such calculations depend in some degree on our relation to the vast aggregate of the works of our Creator; and no person who is accustomed to meditate on these subjects will be surprised that the numbers which such an occasion requires should oppress our comprehension. No one who has dwelt on the thought of a universal Creator and Preserver, will be surprised to find the conviction forced upon the mind by every new train of speculation, that viewed in reference to Him, our space is a point, our time a moment, our millions a handful, our permanence a quick decay.

"Our knowledge of the vast periods, both geological and astronomical, of which we have spoken, is most slight. It is in fact little more than that such periods exist; that the surface of the earth has, at wide intervals of time, undergone great changes in the disposition of land and water, and in the forms of animal life; and that the motions of the heavenly bodies round the sun are effected, though with inconceivable slowness, by a force which must end by deranging them altogether. It would, therefore, be rash to endeavor to establish any analogy between the periods thus disclosed; but we may observe that they *agree* in this, that they reduce all things to the general rule of *finite duration*. As all the geological states of which we find evidence in the present state of the earth, have had their termination, so also the astronomical conditions under which the revolutions of the earth itself proceed, involve the necessity of a future cessation of these revolutions.

"The contemplative person may well be struck by this universal law of the creation. We are in the habit sometimes of contrasting the transient destiny of man with the permanence of the forests, the mountains, the ocean — with the unwearied circuit of the sun. But this contrast is a delusion of our own imagination; the difference is after all but one of degree. The forest tree endures for its centuries, and then decays; the mountains crumble and change, and perhaps subside in some convulsion of nature; the sea retires, and the shore ceases to resound with the 'everlasting' voice of the ocean; such reflections have already crowded upon the mind of the geologist; and it now appears that the courses of the heavens themselves are not exempt from the universal law of decay; that not only the rocks and the mountains, but the sun and moon, have the sentence, 'to end,' stamped upon their foreheads. They enjoy no privilege beyond man except a longer respite. The ephemeron perishes in an hour; man endures for his three-score years and ten; an empire, a nation, numbers its centuries, it may be its thousands of years; the continents and islands which its dominion includes have perhaps their date, as those which preceded them have had; and the very revolutions of the sky, by which centuries are numbered, will at last languish and stand still."

The final obliteration of the system, at least so far as the comets are concerned, was a favorite theory advanced by Newton, and one which, to a certain degree, he seems to have cherished to the latest hours of his life. He thus regarded the comets as "the aliment by which suns are sustained"; and remarked, that although it was impossible for him to conjecture when the comet of 1680 — one to which he had devoted special attention — will fall to the sun, yet, whenever that time shall arrive, the heat of the sun will be raised by it to such a degree, that our earth will be so intensely heated, that all animal and vegetable existence will be destroyed. It has also been urged that the assumption of a resisting medium as resulting in the final destruction of our system, indicates a direct analogy between the laws which govern all parts of the material universe; that perpetual change and perpetual progression, increase and diminution, are found thus to prevail without exception. It must be remarked, however, that this view of the system of the world, although supported by close analogy, does not on that account present unavoidable claims to be regarded as strictly true, or even as being anything more than a mere speculation. It would indeed seem strange and unaccountable that the Omnipotent Creator of the universe should construct systems of worlds revolving around a common centre, in accordance with laws so regulated, if not subjected to external causes, as to perpetuate each individual and compound system. There may be a beginning without an end, although the reverse is impossible, at least so far as human reason can understand; and to conceive, therefore, that because the present state of things has had a beginning, it must necessarily have an end, is by no means admissible. Nor is it enough to say that the resisting medium does not in the least counteract what is most important in the provision for the perpetuity of our system,

and that the order of nature will remain unchanged
for a period, compared with which the age of the
earth in its present state is insignificant. The ultimate effect of the resisting medium will be, as
already noticed, to diminish not only the periods of
the revolutions of the different planets and comets,
but also that of the earth, and thus to derange those
adaptations which depend on the length of the year,
and fit it for the abode of animal life such as now
exists. It may, however, have been preordained
that our own system, at least, should exist for a
specific purpose, and that, having fulfilled its purpose, it should, as such, have its end of duration,
though millions of millions of years were designated
by the Creator as the period in which all its successive transformations should be effected. When
the present order of things had its beginning is
beyond the power of the human reason to declare;
and when, if ever, it shall end, is equally beyond
our comprehension. In all such inquiries we are
soon lost and bewildered by the uncertainties which
surround us, and entering, as we do, the confines
of the unknown, it becomes as impossible to proceed
with the investigation as to attempt to conceive of
the origin of Omnipotence itself. The power of
the human reason is finite, and, therefore, unequal
to the comprehension of the infinite; and, in view
of this, we can only hope that the time will come,

and at no very distant day, when some of the mysteries in which many of the phenomena of the material world are involved, will be finally dissipated, and when the field of speculation, which, to a limited extent, must ever remain, will be finally compressed within very narrow limits.

In regard to the theory, that there exist evident indications that all must end, it might be added that the limit of even millions of millions of ages is by no means sufficient to dispel any tendency in the mind to revolt against any such assumption. Granting, therefore, the existence of a resisting medium in space, and in view of the ultimate effect which would thus be produced, if not counteracted, we are naturally inclined to look to the operation of some other force, yet unknown, to obviate, or, at least, modify the nature and degree of the effect, and thus avert the dreadful calamity which must eventually be produced. The perturbations of the comets on each other are yet undetermined, and also the ultimate effect of the attractions of these bodies on the solar system. The phenomena of the tails of comets have indicated, beyond all doubt, the existence of a repulsive force in all material bodies, while the fact that the force of gravity, and the undulations of the ethereal fluid in producing the intensity of light, vary according to the same law, taken in connection with the supposed — and, we believe, clearly established — identity of light, heat, and electricity, affords cogent reasons for conjecturing that there may exist some unknown bond connecting all these forces of nature in such a manner that their combined effect will be productive of the most harmonious results. In this way it may be considered possible that future investigations will establish such relations between the operation of the various physical forces, as will exhibit in the clearest and most instructive manner, the evident design of perpetuity which the Creator may have enstamped on every part of his universe of creation, already ended.

There is also another point of view from which the conditions of the perpetuity of the systems of the universe may be considered, which must not be passed by unnoticed, and which possibly may furnish an explanation, by no means improbable, of the object of the comets in the economy of nature. If we suppose, that, in the course of an indefinite number of ages, an almost infinite number of these bodies shall pass through our system, it may be concluded that, assuming their individual masses or weights to be small, their disturbing influence in the system will be sufficient to counterbalance any minute, or, so to speak, *residual* inequalities or irregularities which may exist. We may suppose from analogy — and there can, in this case, be no

forcible objection to such a course of reasoning —
that, in the other systems of the universe, there
exist the same conditions and arrangements for
preserving their stability as in the case of our own
system. From what we know of the motions of
the comets, after they cease to be visible from the
earth and in the remote regions of space, it may be
inferred that they pass through, and present in other
systems the same phenomena, only to a greater or
less extent, which have characterized their appearance in our own; and that although some of them
will be drawn for a time, at least, into orbits which
would retain them as members of the system, yet
by far the greater number would move in orbits
which would conduct them beyond the confines of
the various systems which they may visit; and so,
passing on in a meandering course, from sun to
sun, and from system to system, they may poise
even by their infinitesimal masses, as circumstances
may require, every minute irregularity in the systems which they visit, thus preserving, through
time and eternity, the stability and equilibrium of
the universe.

. . . . . .

. . . . . .

Such is the present state of cometary astronomy,
divested, so far as practicable, of its technicalities,
and brought within the comprehension of every

general reader. The phenomena which have been presented are such as furnish the most sublime objects for contemplation; and in what has been stated, whether on the solid basis of established truth, or in the form of a speculation, we have endeavored to be careful to avoid everything which might, should the train of thought be continued further, lead to erroneous conclusions. We have also endeavored to give all the most important facts which the researches of astronomers have developed in connection with the theory of these wonderful bodies, and have shown conclusively, as must be admitted, that these objects have no direct influence in the local affairs of our earth. We have not alluded to any supposed possibility of the comets being regarded as the abode of animal or vegetable life, except in the statement that Whiston supposed these bodies to be the residence of the damned, who are punished by being alternately wheeled into regions of intolerable heat and of the most intense cold. It should not, however, be supposed that, among the various speculations respecting the character of these bodies, they have never been regarded as the abode of rational and intelligent beings. Lambert, in contemplating the excursions which these bodies make in the immensity of space, concludes with the following reflections, which must necessarily be regarded as simply a fanciful conception of a brilliant imagination :

"I love to figure to myself those travelling globes, peopled with astronomers, who are stationed there for the express purpose of contemplating nature on a large, as we contemplate it on a small, scale. Their movable observatory cruising from sun to sun, carries them in succession through every different point of view, places them in a situation to survey all, to determine the position and motion of each star, to measure the orbits of the planets and comets which revolve around them, to observe how particular are resolved into general laws, in one word, to get acquainted with the whole as well as the detail. We may suppose that their year is measured by the length of their route from one sun to another. Winter falls in the middle of their journey; each passage of a perihelion is the return of summer; each introduction to a new world is the revival of spring; and the period of quitting it is the beginning of their autumn. The place of their abode is accommodated to all their distances from the fixed stars, and the different degrees of their heat make the fruits and vegetables designed for their use blossom and ripen. Happy intelligences, how excellent must be the frame of your nature! Myriads of ages pass away with you like so many days with the inhabitants of the earth.

31

Our largest measurements are your infinitely small quantities; our millions the elements of your arithmetic; we breathe but a moment; our lot is error and death, yours science and immortality. All this is agreeable to the analogy of the works of creation. The frame of the universe furnishes matter of contemplation as a whole as well as in each of its parts. There is not a point that does not merit our observation; this magnificent fabric is portioned out in detached parts to created beings; but it is in the unity of the whole that sovereign perfection shines; and can we suppose that this whole has no observers? The imagination, indeed, after so sublime a flight, may be astonished at its own temerity; but, in short, here the cause is proportioned to the effect, and there is nothing great or small in immensity and eternity."

Whether the comets are in a state capable of being the abode of rational intelligences, or whether, as is more probable, they are only exhibitions of matter in a state of transformation, are questions which we shall not consider. They are by no means necessary to a complete understanding of our subject, and for this reason have been omitted. We may, therefore, remark in conclusion, that enough has been presented to almost overwhelm the mind in the contemplation of the material universe.

General Remarks.

"Well hath the great Creator of the world,
Fram'd it in that exact and perfect form,
That by itself unmovable might stand,
Supported only by his providence.
Well hath his powerful wisdom ordered
The, in nature, disagreeing elements,
That all affecting their peculiar place,
Maintain the conservation of the whole.
Well hath he taught the swelling ocean
To know its bounds, lest in luxurious pride
He should insult upon the conquer'd land:
Well hath he plac'd those torches in the heav'ns,
To give light to our else darken'd eyes:
The crystal windows through which our soul,
Looking upon the world's most beauteous face,
Is blest with sight and knowledge of his works."

THE END.